Mastering AutoCAD AEC

R. Shepherd

 LONGMAN

Addison Wesley Longman Limited
Edinburgh Gate, Harlow
Essex CM20 2JE, England
and Associated Companies throughout the world

First published 1996

Trademarks
Throughout this book trademarked names are used. Rather than put a
trademark symbol with every occurrence of a trademarked name, we state that
we are using the names purely in an editorial fashion and to the benefit of the
trademark owner, with no intention of infringement of the trademark.

British Library Cataloguing in Publication Data
A catalogue entry for this title is available from the British Library

ISBN 0-582-09927-7

Set in 10/13 pt Melior
Produced by Longman Singapore Publishers (Pte) Ltd.
Printed in Singapore

Contents

Acknowledgements

The author would like to thank the following individuals:

Colin Priestley of Wm Morrisons Supermarkets, Bradford, for the inspiration needed to write this book.

David Clarke, Simon Jones and the team at Autodesk, Guildford, for writing AutoCAD AEC.

Pat for editing.

MicroCad of Bradford and Ravencad of Chichester for software.

Research Machines for providing the hardware that never failed.

Calcomp and Ricoh for laser plotted and printed hardcopy.

Alan Eden, Michael Hellawell and Associates, Chartered Architects, Bradford.

CHAPTER 1

Introduction to AutoCAD AEC

AutoCAD users and potential users are well served with books on the world's best selling CAD package, even though many of them struggle on with excellent self tutors written in the USA that still use the Imperial units of measurement now only used in a few countries.

AutoCAD AEC users have not been so fortunate however, due to the more specialist nature of the software and the fact that the program originated in the UK where very few computer books are commissioned and published.

1.1 Aims of the work

The aim of this work is therefore to provide AEC users, potential users and students of architecture, building and civil engineering with a work that introduces some of the many features found in such a large and complex software package. Aimed at the professional designer it assumes an investment in equipment that will be matched by a committed learning process. With such a large piece of software and very limited space the projects described in this work are typical of domestic and small industrial buildings. It is hoped that the more complex modelling and structural features of the package will be covered in a later volume. A methodology is established however which leads the active reader/user through a series of specimen design and drawing exercises. Another aim is to give the reader an insight into some of the future possibilities of using an two- and three-dimensional building design package on a personal computer.

There are several reasons for using AutoCAD AEC for construction design rather than standard AutoCAD which revolve in large part around the following areas:

- Increased productivity due to the automation of a number of repetitive routines included within the AEC extension of AutoCAD.

- Provision of many new commands and design routines that can be applied to specialist areas of construction and site design.
- Introduction of new features that extend the capability of a CAD package into other application areas.
- Rationalisation of the AutoCAD command structure for easier use by building design staff who may not be computer literate.

This extension of features in AutoCAD AEC increases the complexity of the software above that of AutoCAD and the user will need time and imagination to develop this fully while acquiring the skills needed to master the package. Proper training in the use of AutoCAD AEC may be neglected if this fact is not recognised, and designers and their managers may think that staff sent on a basic AutoCAD course can cope with the enhanced features of AEC. The adoption of this approach is to misunderstand the design potential of AEC as opposed to the drawing potential of AutoCAD. Using AEC you can create complete elements of structure and manipulate spaces in addition to creating line drawings and surface meshes.

Potential readers

This work is essential reading for students who are studying design technology and processes for degrees in architecture, building surveying, building and civil engineering.

For those architectural assistants and technologists taking the HNC/D Design Technology, Design Procedures and Project units, completion of the project exercises will enhance their skills considerably. Incorporated civil and structural engineers will find the work useful. The book is also intended to supplement the project work for the City & Guilds course 4351-06: AEC in the Built Environment. This volume could also be used by:

- Practising architects, assistants and technologists.
- Practising interior designers.
- Design and build contractors.
- CAD managers and others responsible for the introduction of new technology.
- Estate developers and house builders.
- House purchasers, building societies and banks, and members of the general public interested in building design and drawing.

Note   Potential construction clients, particularly corporate clients, can now participate interactively in the design of a building as it is developed on screen in real time. The time saving achievable in this scenario will be readily appreciated by all familiar with the traditional method of developing alternative proposals on paper only for the next round of functional requirements to trigger changes in design. Clients can now demand from their design consultants a complete three-dimensional (3D) model of the project while they wait. Indeed clients should also be able to see the cost implications on screen as the floor area and other features with cost implications are changed to suit their demands. Fitted kitchens are now designed and approved using 3D packages by many firms; British Gas offers a fully rendered and animated visualisation service using Silicon Graphics machines.

1.2 The project approach

Most of the commands available to AEC users are covered as they are used in a series of specimen design projects. Each project is different in nature and design method to prevent repetition of command usage. Where several design philosophies and methods exist in areas such as elevation generation or 3D modelling the alternatives are set out in separate sections. Usage of various forms of textural information and hard copy output are equally applicable to all of the projects described and have their own chapters.

Project order

1 Small commercial building which could form a garage, shop or single-storey office with a flat roof.
2 Detached or semi-detached bungalow with single or double garage, bay windows and a pitched roof.
3 Elevations and sections are generated from the first two project models.
4 Multi-storey domestic buildings with internal floors and stairs are then created.
5 All the project models are enhanced and blocked, and used to create a site plan and 3D site model.
6 Finally, roads and other site features are added.

More advanced 3D work such as the creation of complex multi-storey buildings and structural elements are not covered in this volume.

It should not be inferred that larger more complex buildings are not suitable design models. Indeed, the larger the building being designed, the greater will be the productivity gains using a CAD package, but they will use specialist techniques outside the scope of this introductory volume.

Many of the standard AutoCAD editing techniques like Copy, Array and Mirror can be used to repeat the details designed for small buildings on a larger scale and thus scale up the scope of the package. As building size increases, it will be the hardware limitations that start to slow down performance owing to the sheer size of the drawing files generated, rather than the software's capabilities.

Note   Symbol libraries are essential tools for enhancing 2D views and the RIBACAD 2D electronic library contains drawings from many manufacturers which are now available to subscribers. These provide the means to quickly add details to plans, elevations and sections needed for working drawings. Standard specifications can now also be associated with many of the symbols included in the RIBACAD library.

Division of the work

The later part of this Chapter is devoted to the equipment required for architectural CAD and an explanation of the basic MS-DOS commands needed to get AutoCAD AEC up and running, followed by a section on installing the RIBACAD electronic library. However, most users will find that the time spent learning to use a DOS or Windows shell such as PC Tools will be quickly repaid by enhanced productivity, everyday file manipulation and directory organisation. Backing-up routines are also included in most DOS utility programs so that when your hard disk fails (as they all do eventually) your data is not lost for good.

Lesson form and explanatory method

Each section of the book follows a common form so that it can be used as a self tutor or a course book to supplement taught work or just to show the casual reader what would be involved in design and drawing with AutoCAD AEC.

The serious reader is assumed to have a copy of AutoCAD AEC and suitable hardware (see later). The instruction sequences contain the following elements which should be noted by the reader before starting. All the drawings created by completing the projects will fit onto one blank high density floppy disk.

• Short introductory section setting out the results expected from the actions which follow.

- Actions to be undertaken are given sequentially in a numbered list (1, 2, 3, etc.).
- Command access is described by reference to both the tablet (digitiser) and pull-down menus. Commands are always shown in bold type to distinguish them from other text.
- Screen prompts are shown in normal text, and the default option is always shown between brackets < > at the end of each line.
- Notes on commands used and related information are given following a *Comment* heading.

Note Finished drawings, where shown, are not intended to represent the standard achievable by an expert user of the system or to show the software's full potential when used by a specialist with many years' experience.

They do show results that can be achieved first time by novice users who follows the detailed sequences described, after only a few hours working on the system without constant general access to AEC software or a computer. They are included as typical construction types and not intended as design ideals or necessarily in full compliance with current Building Regulations.

1.3 Common difficulties

Most new AEC users like most AutoCAD users find the learning curve very steep in the early stages and have difficulty changing their previous manual approach. However, there is certainly no need for the AEC user to do more than an introductory course on AutoCAD before tackling AEC, and only very inexperienced non-vocational users find difficulty with the special awareness needed to navigate the drawing editor.

Any new user's rate of progress will vary but will depend very much on their knowledge of building construction which is assumed to be vocational in nature and extensive enough to understand the terms used in the book.

Some or all of the following problems are normal with most new AutoCAD AEC users so don't be discouraged by a few setbacks.

- The concept of 3D modelling on a VDU is still new to many building designers and seems very strange to some new and potential users.
- AutoCAD AEC is very particular about the way in which a cavity or single wall is created. Any mistake in the design process here and windows, doors and other elements cannot be

inserted properly and the design process must start again when the walls have been redrawn. Creating the openings first and then adding the walls later between openings is a feature that should be included in AEC.

- Walls cannot be started at one height and finished at another. Only the wall tops can be inclined in user-defined sections.
- Flexibility in the routines for insertion of windows and doors is very limited as only one insertion can be made in any one level of wall. No window insertion in a gable end wall is possible, for example where windows are often needed.
- The need to perform a wall convert after each small change to the design to remove lines above and below window and door openings is an irritation. The window or door opening actually extends to the full wall height and is then patched to fill in the gaps with 3D faces not actual elements.
- Alignment and justification of walls is complex and must be preset. No editing is possible after construction.
- Forgetting just how productive an AEC operation is and that AEC cavity walls are very rapidly created in a short time can lead to frustration if operations need repeating.

Note Version 5.0 of AutoCAD AEC overcomes most of the difficulties.

AEC productivity

General tips before starting your journey through mastering AEC are:

- List all the levels set and full details of elements such as wall heights. Keep this list at your side at all times for quick reference.
- Two minutes spent on setting up Status, Grid and Snap settings before starting to draw can save hours of adjustment later.
- Keep the form of the finished project in mind at all times and avoid getting side-tracked by the random creative ideas that CAD so easily generates. Refer to the finished work shown at the end of each section first.

Approach to active sessions

Follow the sequences necessary to produce 2D and 3D examples similar to those described in the text and check your screen result against the screen representations shown at regular intervals.

Do not slavishly follow the sizes and names given in each

example. Use your own requirements and ideas to vary the examples given in the book.

Take a break after each structural stage has been completed, and complete the additional tasks inserted at regular intervals throughout the book to reinforce the sections already covered.

Do not be afraid of making mistakes. The AutoCAD Edit commands Oops, Undo, and Redo give ample opportunity to reverse mistakes and start again at any point in the design or drawing process, even returning to a clear screen if necessary.

Do not draw one line on top of another to try and correct mistakes. This is a common error of manual drafters changing to CAD and will make the drawing difficult to edit at a later stage.

Remember that a little time spent on setting up basic parameters can save hours of back-tracking later, so always have in mind the drawing aids that are not plotted but which are invaluable guides to rapid drafting: Grid, Snap, Layer, Line type, Scale, Sheet size, Level, Colour, etc. Think of this as the equivalent of planning a drawing sheet to accept different orthographic projections correctly positioned and spaced.

Keep that previously mentioned sketch pad handy so that you can work out ideas in sketch form on paper or check that the dimensions envisaged will work when used on the finished drawings.

Remember to use the AutoCAD enquiry commands frequently to determine the current status of all elements in your drawing. Measure, Status, List, Distance, etc. are all useful ways of determining the precise relationship of one element with another.

Learning a CAD package is a unique experience, unlike most software where well-defined procedural steps must be taken to achieve a predictable end result. With CAD you can adopt many methods to produce the same end result. Your way may not even have been thought about by the software manufacturer, but it can still be worthwhile: such are the endless possibilities of computer graphics.

On many occasions several alternative methods of constructing building elements will be possible. Just check the number of steps involved and the time taken by each alternative method before settling on a definitive routine.

Note Remember that all newcomers to CAD spend an initial period at the bottom of the learning curve, when nothing seems to work and no useful production is achieved. To minimise this period, it is worth starting an AEC evening course or getting a few hours per week on a computer system with AEC installed and using this book as a reinforcement to your course rather than as a complete self tutor. Then, when the crunch comes, you will be productive from day one when starting a real project for the first time.

1.4 Essential equipment

Desktop for CAD

You must be comfortable even if you are left handed and play golf, so do not let anyone sell you a fixed desktop. Just buy a plain table with plenty of metal bracing and a space for cabling to hang down the back and you can arrange the desktop 'as you like it'. Remember that you will want to use a sketch pad, access brochures and view letters with colleagues in small meetings at the table so make sure that the keyboard can be tucked away where the keys cannot be accidentally depressed during a presentation or discussion. It is a good idea to provide a felt lining to the base of the tablet's puck and a clean cubby-hole to tuck it away in, or, if it's cordless, keep it with you when you leave your workstation. Right- and left-handed desktops are illustrated in Fig. 1.1.

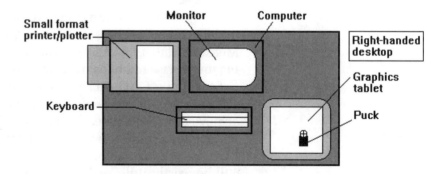

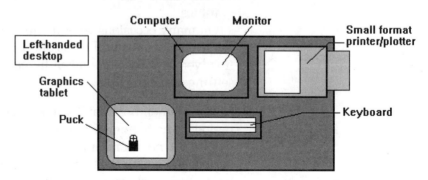

Fig. 1.1

Hardware requirements

Computer

Purchase whichever of the available machines that has the best price/performance ratio at the time of purchase. Your most suitable machine will probably be based on a one or more of the processors mentioned below. Major manufacturers offer units based on the Intel (486 or Pentium), Motorola, IBM, Cyrix, AMD, PowerPC, NextGen, Mips or Alpha chips. Lack of consumer choice will never be a problem in computer purchase now that the Intel monopoly in processor supply has been broken, but making the right decision may be a lengthy process. Always remember that the software requirements govern hardware purchases and should always be the starting point of any CAD system.

Monitor

The preferred monitor for 3D modelling in several viewports is a 21" screen displaying at 1280 × 1024 resolution. If your main work is 2D and you want the best price/performance ratio go for the 17" monitor displaying at 1024 × 768. It is not worth considering 14" or 15" monitors as they offer no price/performance advantage over the 17" variety.

Graphics card

Overnight changes are the rule with graphics card development and many CAD users have bought expensive systems which have been top performers for a few months, only to vanish into obsolescence. Advances in rendering, virtual reality and video capture will be a constant spur to the production of faster hardware and to the chip makers to respond with faster processor performance. Try to spot the emerging standards and the higher levels of support offered by groups of different companies. Avoid standards that originate from one company.

Command input devices

The keyboard and mouse are unlikely to alter much in function although styles may change, but graphics tablets (digitisers) are still improving and many new features are evolving at ever lower costs. If a method of projecting the graphics tablet surface from the horizontal to the vertical could be found then they could provide all the facilities of on-screen icons without reducing the on-screen drawing space. They also have a vital role in protecting CAD users eyesight by forcing the eye to refocus at regular intervals, so for several reasons their use is highly recommended.

In this volume the use of mouse selections from pull-down menus to input commands is fully covered and selections from the digitising tablet by puck are also described for each command. The best graphics tablets (digitisers) can also be configured as multi-mode mice. Many commands can be typed in from the keyboard but most architectural users will probably not have the skills to gain the rapid command input that skilled use of the keyboard can bring.

Hardcopy devices

The trusty large-format vector-based pen plotter, for so long the mainstay of CAD hardcopy output, is now becoming obsolete, and the raster-based inkjet printer is taking over its role as the low-cost large-format output device. The range of inkjet machines available is very large and have wide price differentials based on features such as mono (black) or colour output. Even machines capable of photographic quality output are relatively cheap compared to their electrostatic predecessors. Direct imaging and laser-based machines are also making rapid technical progress at lower production costs. (See Chapter 8 on hardcopy output).

Software requirements

- AutoCAD Release 12 or later for DOS or Windows.
- AutoCAD AEC 3.0 or later for DOS or Windows.
- OS/2 or DOS 5.0 or later and Windows 3.0 or later.
- Anti-virus software.
- Accurender, ARE 24 or other rendering package.
- Drawing management software such as Slick or Cyco.
- Display list software such as Panacea or SoftEngine.

Choice of operating system   Versions of AutoCAD Release 12 or 13 and AEC 4.0 or 5.0 are available running under different operating systems. The functionality of each version remains virtually the same whichever operating system is used however, so a decision on which version to purchase depends entirely on your type of work and the interaction you need with other application packages that you use.

If you are intending to produce technical manuals on A4 paper then you will need to mix text, graphics and numeric information frequently throughout the document. Under these conditions the OLE and DDE functions of Windows, together with the ability to open multiples drawings and documents at the same time and to

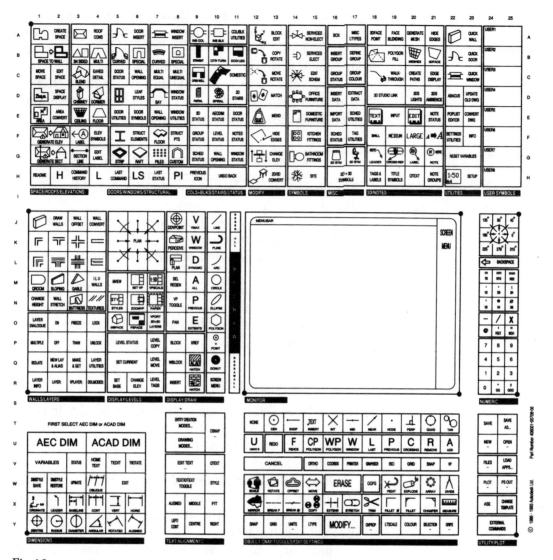

Fig. 1.2

cut, copy and paste between them, will be most beneficial to your output.

However, if the majority of your work is in producing large-format drawings that include text and graphic design or working details you will find that the Windows interface is unacceptably slow for most projects. In these circumstances the 32-bit DOS extended version of AEC will give a three- to four-fold increase in speed and even then, on occasions, will not keep up with your thought processes or manual dexterity. Waiting around for Windows-based applications to respond is not a very satisfying experience. The use of a graphics tablet overlay, such as the

AutoCAD AEC 4.0 overlay shown in Fig. 1.2, is recommended for architectural users with the DOS version of AEC. Windows users will probably prefer to use a mouse as their version of AEC is designed for this method of command input.

New versions of AutoCAD and AEC will include as standard specialist graphics software for much better manipulation of 3D models on screen and faster rendering.

Rules for hardware and software purchase

Try to anticipate developments in technology 12 to 24 months ahead.

Spend more than you think you can afford at the time to buy the best specification available.

Make reliability a first priority and purchase cost a secondary factor.

Pick and mix components to get the best. Use the services of a specialist systems integrator to build up a custom system. Using only one vendors complete system restricts your choice and is usually more expensive.

1.5 Basic DOS

The only reason that this section is included is to help enable the AEC user to manage the large volume of RIBACAD library drawings that are mentioned & used in later sections to enhance 2D elevations and sections.

Observations are based on the reactions of designers and draughtspersons new to computers faced with a hostile unfriendly system that manages the drawing environment but which seems unrelated to it at first.

However once the section on use of the RIBACAD library is reached the importance of good directory management becomes obvious.

For instance, use the AutoCAD Insert block command to insert RIBACAD drawing TH110Y. In using the Blocks menu a prompt appears – Block name: or ?: – typing in A:TH110Y is much easier than typing C:\RIBACAD\TH\TH110Y though this implies that you know how to copy a drawing file from the hard drive to a floppy disk in the first place as the RIBACAD files are installed on a hard disk.

Whilst considering file management it is worth noting that there are now many utility 'shell' programmes available to help the CAD user organise the working environment, and most try to give a point-and-shoot or drag-and-drop facility on screen so that the options available are directly presented to the user.

Look at the file management features found in Windows 3.1, DeskView, PC Tools, Lotus Magellan, XTREE, Synergy, Norton and as many more as time permits; they all help you manage DOS better and quicker so that drawing time is maximised and the time spent fine tuning the operating system *minimised*.

Refer to the help and tutorial sections of Windows 3.1 and the other programmes mentioned above and see how the execution of the DOS commands described in this section can be quickened.

Browsing through drawings to find the one you forgot to number can also be a pain so consider purchase of a drawings file manager such as Slick, Cyco or one of many others available which will display and manage your drawing files quickly.

Most if not all of the DOS commands considered later are both easier and quicker under the DOS shell programmes mentioned above, and the directory tree shown here is similar to the one that appears in many file management programmes.

Although AutoCAD is fairly easy to install, AEC adds some complexity and 15 minutes to the installation process so get the dealer who supplied the package to install your first AEC system and set up a suitable batch file such as the one shown below.

Some of the functions mentioned under the basic DOS commands can be accomplished more easily using the AutoCAD Utility menu (see the AutoCAD manual for full details). Others are mentioned in relevant sections of drawing practice.

Assuming that AEC has been successfully installed following the comprehensive instructions given in the installation manual the management of RIBACAD files can be used to illustrate many useful DOS commands (see Fig. 1.3).

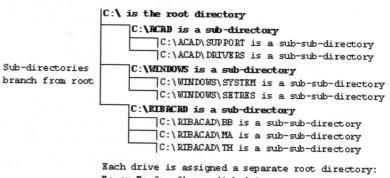

Fig. 1.3

Process stages

1   Unpack your floppy disks which will contain all the drawing files in a compressed form.
2   Check that command line prompt reads C:\> and that the cursor appears just after the prompt. Note that all commands must be entered with spaces as shown.
3   Make a directory on your hard disk in which to install the drawing files:

command: **MKDIR C:\RIBACAD** enter

Comment   This command should be used only if you do not want to place the RIBACAD files in the root directory. In general terms files are accessed faster if split into sub directories but it means more typing when you need to insert them into drawings. Note that *enter* indicates throughout this book that the enter or return key should be pressed.

4   Change to the directory just made:

command: **CHDIR C:\RIBACAD** enter

5   Install the drawing files within the directory just created:

command: **A:INSTALL** enter

Comment   You may need to press the space bar and type A: to activate the floppy drive. If you have not used the CHDIR command prior to installation all the files will be placed in manufacturer's sub-directories next to the root directory as shown in the directory tree.

6   Just insert a new disk when prompted after removing the one whose files have been installed.
7   When installation is complete various cryptic messages about file decompression will appear on screen.
Comment   RIBACAD now uses a new and much better system of decompressing files which takes minutes not hours to extract all the files on a floppy.

8   Check to see if the files have been installed on the hard disk:

command: **DIR** enter

Comment   A long list of manufacturer's sub-directories will scroll on screen: check them against documentation.

9   Check each sub directory to see if files installed correctly:

command: **CD \RIBACAD\BB** enter
command: **DIR** enter
or
command: **DIR /P** enter to view one page of files at a time

Comment   Individual file names such as BB032F and BB037F will now appear on screen followed by the size of each drawing in number of bytes.

10   Copy the files in each sub directory to high density floppy disks for storage:

command: **COPY \ BB A:** *enter*

Comment   This command copies all files in the \BB sub-directory onto floppy disk. This is where the DOS shells mentioned previously are really useful as they enable files to be compressed by 50% and manipulation of tagged blocks of files or individual files is possible. (To copy individual files from the hard disk to a floppy a command such as **COPY \ BB \ BB032F A:** could be used.)

11   Delete the RIBACAD files from the hard disk to save valuable space on the hard disk which may be needed as extended memory or used by drawing files during a hide command.

command: **DEL \ BB** *enter*
prompt: Are you sure (Y/N)?
type: **Y** (for yes) *enter*
command: **DIR** *enter*

Comment   The screen should show only two files left in the directory (. and ..) which are not data files but system files needed by DOS. (To delete individual files within the \BB directory they should be specified separately.)

12   Now the directory can be removed from the hard disk but not whilst in the current directory so change to another directory first:

command: **CD C:** *enter*

13   Then remove the directory which is no longer needed:

command: **RMDIR C:\ RIBACAD\ BB** *enter*

Comment   The directory and its . and .. files will be removed by this command.

Summary

After this practice you should be able to install programme or library files on the hard disk, check which files have been installed and how much disk space is occupied, copy files from one drive to another, delete files and remove an empty directory from the hard disk.

1.6 Setting up for project work

Starting the programme

Select the batch file that starts AEC and inserts a default drawing editor (see Fig. 1.4). The batch file can also have parameters set for automatically inserting on screen a standard hard copy style which includes scale, size, border type and other parameters.

Fig. 1.4

```
COPY C:\ACAD\ACADL.CXP C:\ACAD\ACADL.EXP
SET ACAD=C:\AEC\A\D;C:\ACAD\SUPPORT;C:\ACAD\FONTS
SET ACADCFG=C:\AEC\A
SET AECVARS=NONE
SET AECMODE=PROMPT
SET ACADDRV=C:\ACAD\DRV
CD\BATCH
C:\ACAD\ACAD %1 %2
```

Each of the projects that you undertake will probably require a selection of hardcopy styles for different types of drawing. A description of some setup styles is therefore included below. In this work there are several projects that will require startup settings of sheet size, scale, border style and other parameters to be made.

Start any AEC project work by selecting the AEC option from the opening screen dialogue box as shown in Fig. 1.5.

The AEC set up dialogue box will appear as shown in Fig. 1.6.

1   Select 'Start NEW drawing' within the AEC dialogue box.
2   Type drawing name at prompt, for example 'SHOW1' or any drawing name with up to 8 characters.

Fig. 1.5

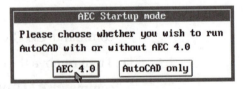

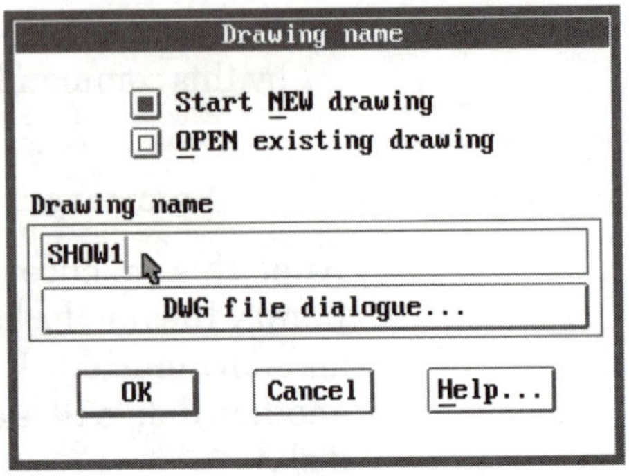

Fig. 1.6

Comment   No spaces are allowed in the drawing name. Space is allowed in the longer names allowed by other operating systems.

There are two approaches to setting up a drawing sheet to start the project:

3   Select a named style from the dialogue box shown below to produce a template for the drawing sheet including details of the project.
   Comment   Picking the option shown in Fig. 1.7 produces the drawing sheet shown in Fig. 1.8 where other details of the drawing and the project can be inserted.
   or

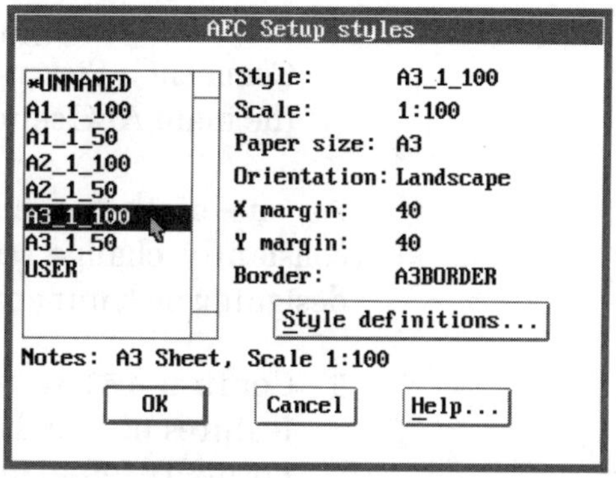

Fig. 1.7

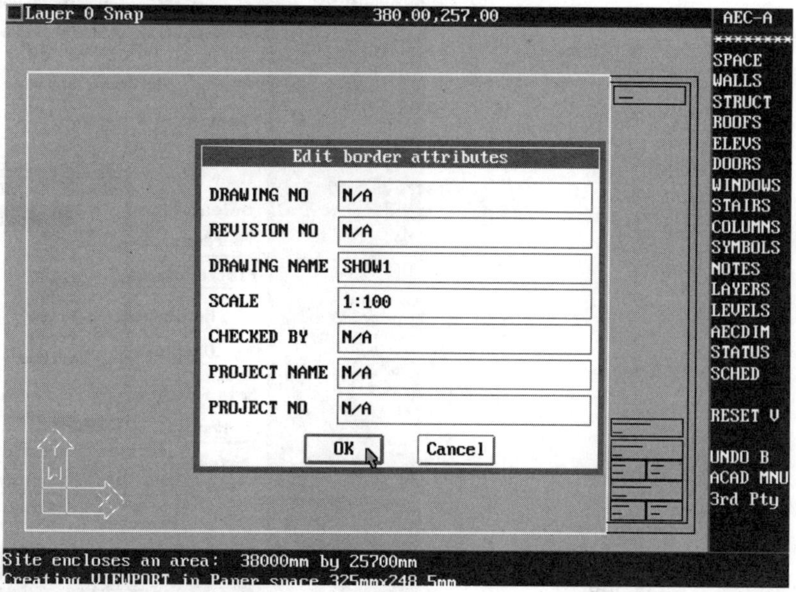

Fig. 1.8

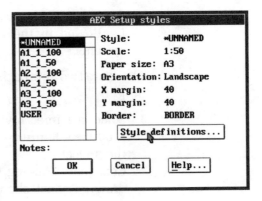

Fig. 1.9

4   Select "UNNAMED' from the style box to access the AEC setup style dialogue box which enables individual choices to be made concerning sheet size, scale and plotter margins.
Comment   Picking unnamed from the style box (Fig. 1.9) calls the main AEC setup menu shown in Fig. 1.10.

As one of the main reasons for using CAD is the ability to constantly change scale sheet size and other parameters while designing pick unnamed as the standard option.

5   Confirm 1:50 as the hardcopy scale and examine the other features of this dialogue box which contains all the elements normally found in the AEC repertoire of dialogue boxes.

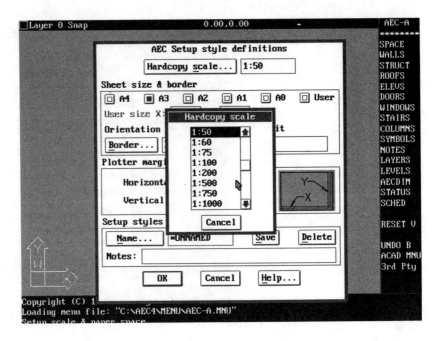

Fig. 1.10

Dialogue box conventions

Many settings can be changed from one dialogue box but the conventions used must be recognised first.

• Solid black dots indicate the chosen setting from a fixed range of buttons.
• A cross in a toggle box indicates that the option is enabled; the box is blank when the option is disabled.
• Greyed out text indicate options activated by a range choice.
• Text boxes contain values that can be changed by selecting the box and typing in a new value.
• Terminators are boxes that conclude the selection process.

Note   Dialogue boxes are faster than text prompts but the opportunities for error are much greater so double check the values entered before acceptance.

AEC Drawing Setup

Select the following settings from the AEC setup style definitions dialogue box (Fig. 1.11).

6   Pick A3 by clicking on the A3 button.
Comment   Check the plotter handbook if unsure about manufacturer's requirements. The plotter margins used will depend on the type of plotter being used and the paper size. If yes is

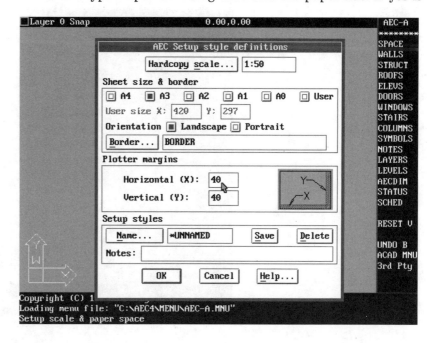

Fig. 1.11

picked from screen menu both X and Y margins can be set and retained as defaults for future plots. Check the paper after a check plot to see width of damage to paper edges and set margins to suit.

7   Pick the Landscape orientation button.

8   Set plotter margins by typing in new values.

9   Click OK after checking settings.

A border will now appear on the screen to represent the edges of a sheet of paper A3 size and the UCS model space icon will also appear on screen.

This is only the start up size however and it can be changed at any time. To do this pick Setup Option on the tablet (H22–23) or from the Settings pull-down menu and change the scale and sheet size during the design process. This option provides a very useful way of controlling the drawing display on the monitor as it can be changed at any time to give more drawing space as the number of views shown increases. A setup file is created each time the drawing parameters are changed.

Up to this point the full drawing screen (or Drawing Editor in AutoCAD parlance) was not accessible and if for any reason you terminate the setup routine it will not become available. Start again from the beginning if this happens. Elements found in the Drawing Editor are shown in Figs 1.12 and 1.13.

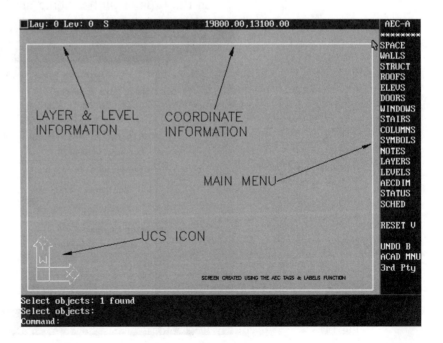

Fig. 1.12

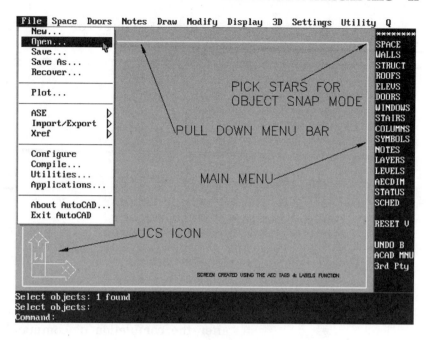

Fig. 1.13

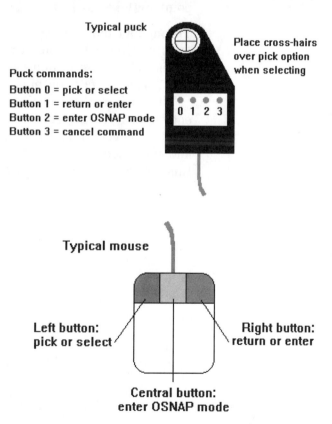

Fig. 1.14

AutoCAD function	Assigned keystrokes
Ctrl	Ctrl
Flip screen	Fl
Toggle coords	F6 or Ctrl + D
Toggle grid	F7 or Ctrl + G
Toggle ortho	F8 or Ctrl + O
Toggle snap	F9 or Ctrl + B
Toggle tablet	F10 or Ctrl + T
Up cursor	↑
Down cursor	↓
Left cursor	←
Right cursor	→
Menu cursor	Ins
Screen cursor	Home
Abort cursor	End
Fast cursor	PgUp
Slow cursor	PdDn

Fig. 1.15

Note   The pull-down menu bar only becomes active before and after the completion of commands when the puck or mouse pointer is taken to the top of the drawing screen.

Throughout the drawing routines the conventions shown in Fig. 1.14 are used for the creation and confirmation of commands.

The left mouse button or the puck's pick button are used to 'pick', 'select' or 'click on' menu and overlay options.

The keyboard enter or return button can be used also where *enter* appears in the text.

Other keyboard shortcuts are listed in the AutoCAD manual and mentioned in the text as needed. Fig. 1.5 shows the use of function keys to toggle drawing settings on and off.

CHAPTER 2

First project: single-storey commercial building

There are many types of project which involve single-storey brick or stone construction with a flat roof and a variety of glazing techniques. The first project will take you through the various processes needed to construct a 3D model or 2D plan and elevations of such a building.

2.1 Building up the space diagram

Space diagram design method

The design work can start by drawing the walls direct or by creating space diagrams first and then converting these to cavity and single walls later. This method is very useful for housing, offices, hotels, and showrooms where the room sizes that are economic to build dominate the design requirements or where minimum space standards are legally specified.

Alteration and improvement work or special projects which are unique are unlikely to benefit from this method as the spaces are more than likely not rectangular or the walls even straight.

Most garages filling stations and car showrooms are built to a standard design with a few alterations to suit the individual site so they are ideal for the space diagram method of design.

Irregular areas in special projects may be built up using the Polyspace command; however, this technique is not covered in this volume.

Composite design method

Another useful technique is to allocate spaces to specific areas of the proposed building and then to trace round those areas to draw the walls in later.

Note   The optional walls compensation feature preserves the full size of the rooms by moving the walls outwards if desired.

If you pick any space definition from the tablet menu drop-down menus appear and now contain the options previously contained in the screen menus which were used in versions 1.0 and 2.0 of AutoCAD AEC.

The construction of irregular shaped spaces is easily accomplished by the software but not included in this introductory work.

Practical

During this session you will practise the following:

- Using the space diagram feature to lay out a small garage floor plan with attached showroom, office, toilets, and any other required spaces.
- Automatic conversion of spaces to walls.
- Viewing the resulting model in 2D and 3D.

Inserting room spaces

To insert a showroom, workshop, office and toilet as individual spaces.

The showroom

1   Pick Create space diagram ... from the Space pull-down menu (Fig. 2.1) or from the tablet [A1–2].

From the dialogue box on screen select the following settings by pointing and clicking or typing in boxes as shown in Fig. 2.2.

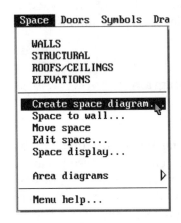

Fig. 2.1

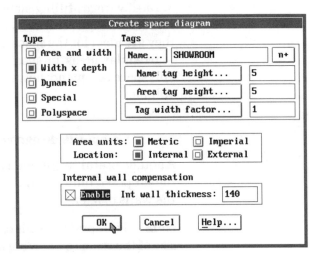

Fig. 2.2

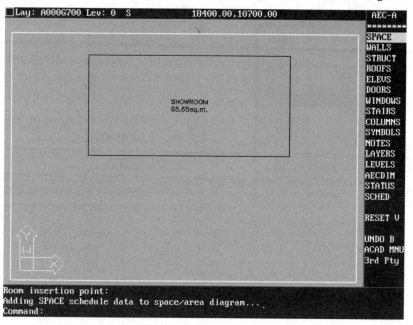

Fig. 2.3

2   Type: Width × depth.
3   Name …: type in SHOWROOM.
4   Internal wall compensation: pick Enable.
5   Int wall thickness: type in 140.
6   Click on OK.

Now answer the command line prompts below the drawing editor.

7   Prompt: Room width: **11500** *enter.*
    Comment   Horizontal dimension on screen.
8   Prompt: Room depth: **5500** *enter.*
    Comment   Vertical dimension on screen.
9   Prompt: Room insertion point: drag onto screen and place
    somewhere near top.
10 Press pick button to fix position on screen (Fig. 2.3).

        The toilet

11 Pick Create space diagram … from the space pull-down menu
    or from the tablet.

From the create space diagram dialogue box on screen select the
following settings by pointing and clicking as shown in Fig. 2.4.

12 Type: Area × width.
13 Name: pick TOILET from the scroll menu.
14 Internal wall compensation: Enable.
15 Int wall thickness: 140.

Fig. 2.4

Now answer the command line prompts below the drawing editor

16 Prompt: Enter the floor area: **9.5** *enter.*
17 Prompt: Enter room width: **2850** *enter.*
18 Prompt: Room insertion point: drag onto screen and place near bottom left corner using the pick button (Fig. 2.5).

Watch carefully the mixture of millimetres (mm) and metres (m) in the area and width prompts.

Office space

You should by now be getting the hang of space insertions so place a third area on screen without any help from me.

19 Place office by using Type: Width × depth.
Comment   The screen should now look like Fig. 2.5. You can attach the rooms directly as you insert them into position on the screen but only if the default OSNAP intersects are correct.

Generally it is better to insert all spaces first and then move them into position later.

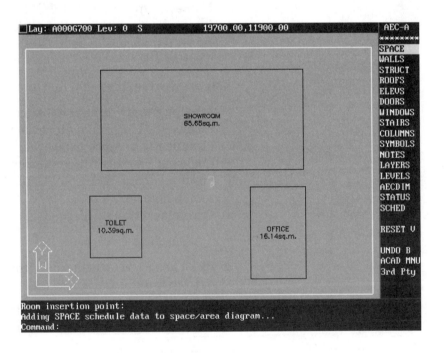

Fig. 2.5

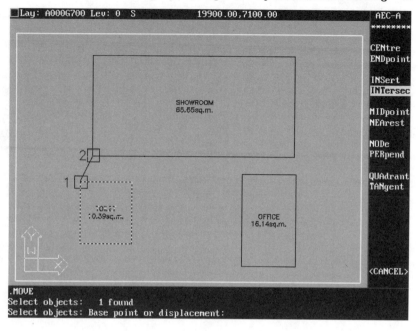

Fig. 2.6

Moving room spaces into final positions

Next move spaces to exact points so that they are in perfect alignment with the other spaces as shown in Fig. 2.6 below. Object snaps must be used when moving spaces to ensure exact alignment otherwise correct wall conversion will not be possible.

1 Pick MOVE from the SPACE on-screen menu, from the tablet [H3] or type in from the keyboard.
2 Prompt: Select objects: pick any point on the space to be moved.
3 Prompt: Base point: pick OSNAP INT from screen menu.
4 Prompt: INT of: pick point 1 (Fig. 2.6).
   Comment   The OSNAP mode is automatically toggled on
5 Prompt: Second point: pick OSNAP INT from screen menu.
6 Prompt: INT of: pick point 2 (Fig. 2.6).
   Comment   Without the use of OSNAP it is impossible to align the spaces exactly.

Now move the Office to join the showroom at the left hand end as shown in Fig. 2.7.
   Next place the workshop into position using the dynamic type option from the create space dialogue box.

7 Prompt: Pick first corner of room: pick point 1 (Fig. 2.7).
8 Prompt: Pick second corner of room: pick point 2 (Fig. 2.7).

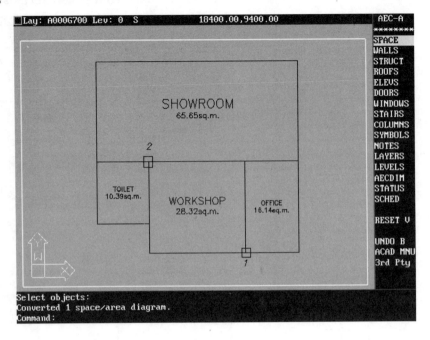

Fig. 2.7

The workshop space is now inserted on screen between the two points picked. External spaces can now be created in AEC 5.0 so this feature is no longer restricted to rooms it can also be used for piazzas.

Note   Any of the many recognised OSNAP points can be used to fix the two corner positions of the space.

Editing the space diagram

Many of the space diagram parameters can be edited by using the space edit dialogue box accessed from the space edit pick on the space pull-down menu. As you can see the tag sizes for the rooms need adjusting to fit each room correctly.

To alter any feature (attribute) of a space:

1 Pick Edit space ... from the Space pull-down menu.
2 Pick any part of the space diagram to activate the Edit space diagram dialogue box (Fig. 2.8).
3 Pick Name tag height ... and type in the new value.
4 Pick Area tag height ... and type in the new value.
5 Change any other value and click OK.

The results of several edit operations are shown in Fig. 2.9.

Note   The size shape and labelling of the space diagrams can be edited by the use of dialogue boxes but the drag and drop methods

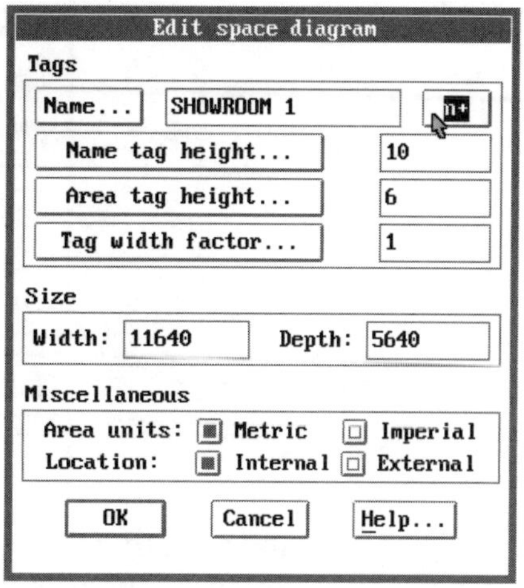

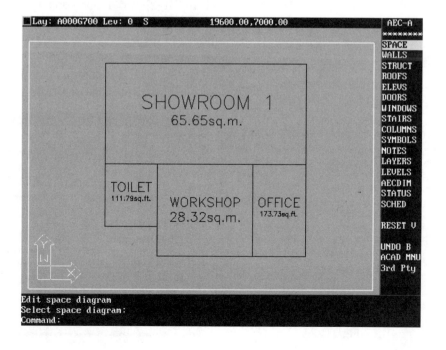

Fig. 2.8

of object oriented graphics packages are still not available. When Windows or other GUI-based versions are available this should enable much more flexible manipulation of the space diagrams. Also note the BS1192 layering convention and the level number on the status bar as these features will be used in later sections.

Unit conversion and area definition

Conversion of the units used can be done in the space edit dialogue box or by picking convert area from the space pull-down menu

Fig. 2.9

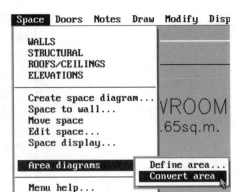

Fig. 2.10

and selecting the spaces to convert (Fig. 2.10). Many other uses exist for space diagrams in more complex forms which are outside the scope of this volume.

Converting space diagrams to walls

Provided the space diagram has been properly constructed conversion to a 2D or 3D walled model is a fast and painless process.

To change the space diagram to a floor plan or 3D model including either single or cavity wall construction:

1   Pick Space to wall ... from the Space pull-down menu or the tablet [B1–2]. A dialogue box appears which gives full access to all internal and external wall parameters that may be preset as shown in Fig. 2.11.
2   Set external wall type to Cavity.
3   Set cavity wall dimensions to values of your choice.
4   Set wall height to 3500 or similar.

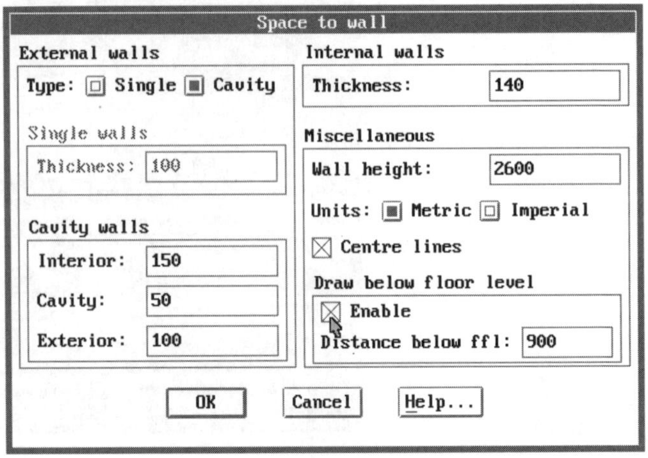

Space to wall	
External walls	**Internal walls**
Type: ☐ Single ■ Cavity	Thickness: 140
Single walls	Miscellaneous
Thickness: 100	Wall height: 2600
	Units: ■ Metric ☐ Imperial
Cavity walls	☒ Centre lines
Interior: 150	Draw below floor level
Cavity: 50	☒ Enable
Exterior: 100	Distance below ffl: 900

OK   Cancel   Help...

Fig. 2.11

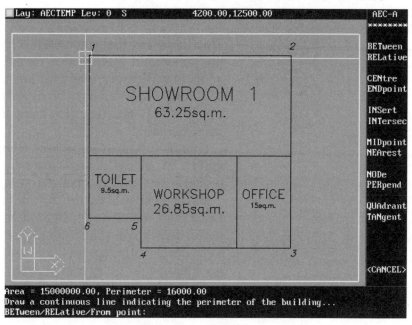

Fig. 2.12

5   Enable footings and set Distance below ffl: to 900.

6   Prompt: Place window round space diagram:

Prompt: First corner: pick first point at top left of diagram.
Prompt: Second corner: pick second point at bottom right of diagram.

7   Select values from the dialogue box which appears:

Interior wall compensation thickness: 140.
Room areas will be adjusted to take away: equal area from each room.

8   Prompt: Draw a continuous line round the building: pick points 1 to 6 as shown in Fig. 2.12.

Note See the following description of space display for the various display options for space diagrams and walls. Note how space is taken equally from each side of the room spaces to form the walls and preserve the original size of the spaces.

If like me you have disabled autosave now is a good time to pick Save as ... from the File pull-down menu or the tablet [T24–25] to save all your drawing work.

Displaying wall and space floor plans

The floor plans that you have just created can now be displayed on the screen and plotted in a variety of ways to represent spaces, walls, or walls and spaces together. This will enable rough cost and structural implications to be determined. They can also be used as a basis for alteration by free hand sketch should the client change his or her mind.

One of the most important features of a CAD system is this ability to change layout very quickly compared to drawings produced in the traditional manner.

Building clients will in future have enough knowledge of the technology to demand instant plan changes and 'what if analysis' performed in real time with all the decision makers watching the same design drawings on their own terminals and able to alter the drawings on the other participants screens.

Three display options are available which can be most easily accessed from the Space pull-down menu.

1   Pick Space display ... from space pull-down menu or the tablet [D1–2] (Fig. 2.13).
2   Pick the required option from the dialogue box that appears (Fig. 2.14).

Note   As a new user you will not want to get involved with the complexities of layers at this stage but all the space display

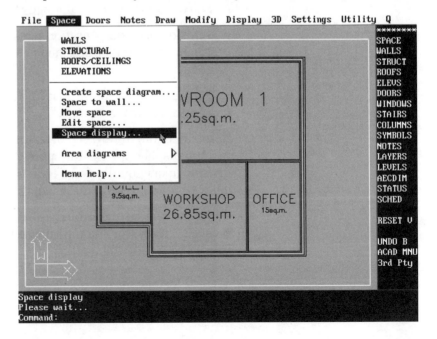

Fig. 2.13

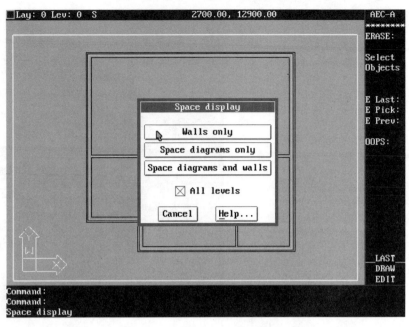

Fig. 2.14

function does is to switch layers on and off. When adding further building elements to the design use the walls only display your default option to avoid any problems with text display.

Pick Space display ... if you need information about room names, areas or perimeters.

3D viewing features

It is now worth viewing the drawn model in 3D by picking either the Dynamic 3D option on the tablet [J8], or the 3D view ..., DDvpoint ... or Vpoint from the 3D view sub-menu of the 3D pull-down menu (Fig. 2.15).

Fig. 2.15

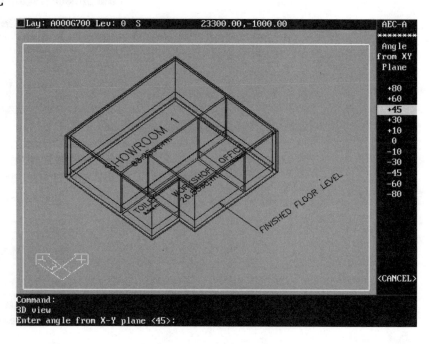

Fig. 2.16

To use the 3D view select a horizontal view point and an angle above the XY plane using the screen menus, as shown in Fig. 2.16.

To use the DDvpoint pick it from the 3D view sub-menu of the 3D pull-down menu. The Viewpoint Presets dialogue box that appears reflects the current view position, as shown in Fig. 2.17.

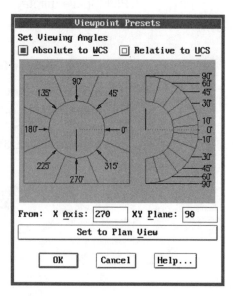

Fig. 2.17

To change the settings just click positions on the screen or type in the new values as shown in Fig. 2.18.

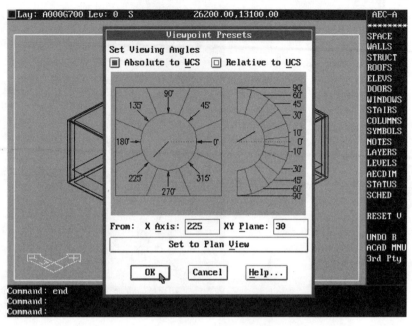

Fig. 2.18

As you can see an isometric view has been set up very quickly by using just two clicks of the mouse or puck.

Note   Make sure that the viewing angle is set as Absolute to WCS *not* Relative to UCS; their magic meanings will be discussed in a later volume.

Picking Vpoint from the 3D view sub-menu will enable you to set up a dynamic view using the bull's eye icon on screen.

The only way to become familiar with the dynamic bull's eye view is to practice moving the small cross around the bull's eye and picking points in each segment. Pay particular attention to view-points from above the 3D model and note the effect of picking the North Polar point on the X, Y and Z coordinates and also picking points where the equator meets the crossed lines. These views represent a plan and an elevation respectively.

Note   See the AutoCAD reference manual or any good AutoCAD book for a full description of the viewing positions available using Vpoint options.

Using the Hide command

Pick Hide from the 3D pull-down menu or type in the commands from the keyboard and watch the command line to see what percentage of the lines have been hidden as shown in Fig. 2.19.

Note   Although the 'hide' on this drawing will only take a few

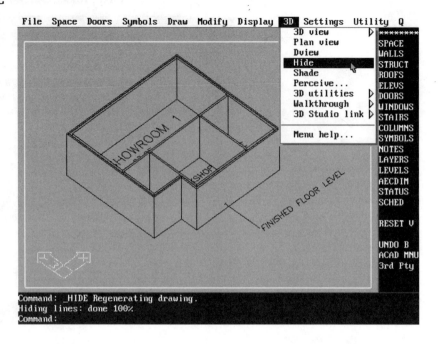

Fig. 2.19

seconds there is good reason to be cautious before hiding a large drawing as it can take some time even with a fast processor. Note that unlike earlier versions of the software the text now obeys the same rules as other drawing elements and is hidden when overlaid by other elements.

Shaded view from the same angle

Pick Shade from the 3D pull-down menu and watch the command line to see what percentage of the lines have been hidden (Fig. 2.20).

Perspective views

Perspective views of the design may also be generated at any time from any viewpoint to produce presentation drawings that are also dimensionally accurate. With an in-built link to Autodesk's 3D Studio these perspective views can be rendered to give photo realistic images or turned into virtual worlds for many purposes. Views are generated by placing viewers anywhere in 3D space and pointing their eyes in a specific direction to give a field of vision. Practice is the only way to become competent in the use of this command which can have an unlimited number of parameters. Actual viewing is achieved by placing a notional viewer anywhere on the plan view at ground level or on any level previously added and then pivoting the viewer horizontally or vertically to change the field of vision.

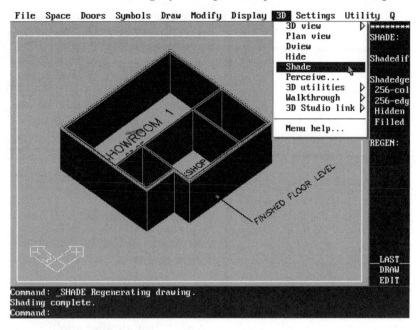

Fig. 2.20

To generate a perspective view of the walls drawn, first size the model if necessary to give more room on screen for the viewers by picking Zoom, Scale X from the Display pull-down or by typing Zoom, picking the Scale X parameter and typing .5 (or another figure) to change the on-screen scaling of the model. Then:

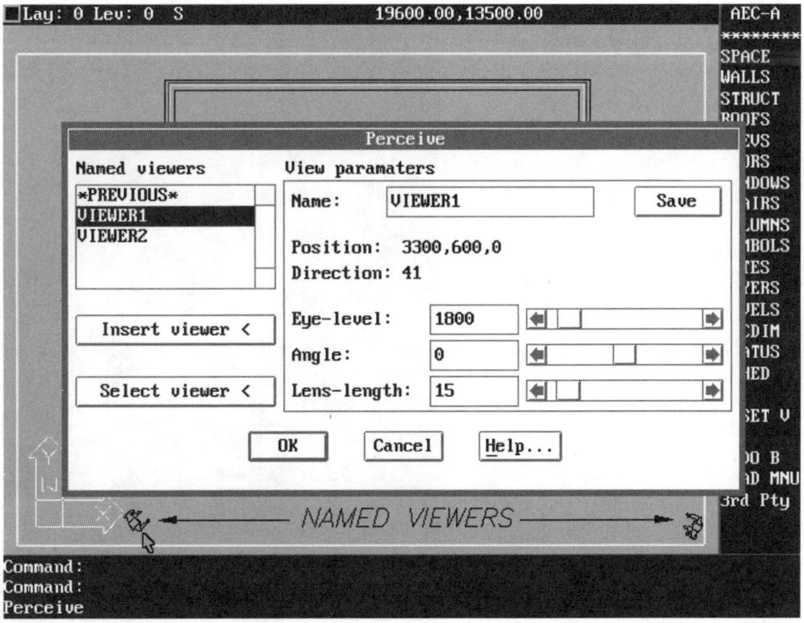

Fig. 2.21

1   Pick the Perceive command from the 3D pull-down menu or the tablet [K8]. This will call the Perceive dialogue box (Fig. 2.21) and enable many settings to be made.
2   Click on Insert viewer.
3   Prompt: Insertion point: pick bottom left corner of screen.
4   Prompt: Rotation angle: rotate the viewer using the 'rubber band' as a pivoting device until the band is aligned to the centre of the viewer's field of vision then press the pick button.

Comment   Pick the Save option in the dialogue box to save viewers for future use. Pick the Select viewer option to use a previously saved viewer .

Viewing options

These are also set up in the perceive dialogue box as shown above in figure perceive.

5   Pick the following options from the dialogue box:

Eye-level: 1800 (or another height)
Angle: 0 (or another angle)
Lens-length: 15 (or higher figure)

6   Click OK and the perspective view will appear on screen.

0 = horizontal view from viewer's eye level
+10 = maximum upward view from viewer's position
−10 = maximum downward view from viewer's position

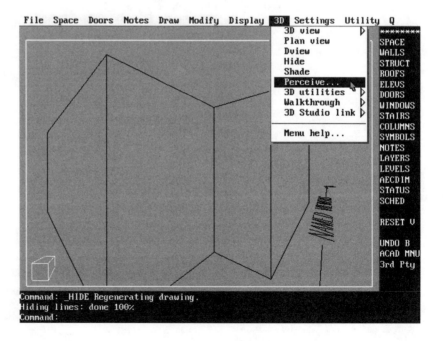

Fig. 2.22

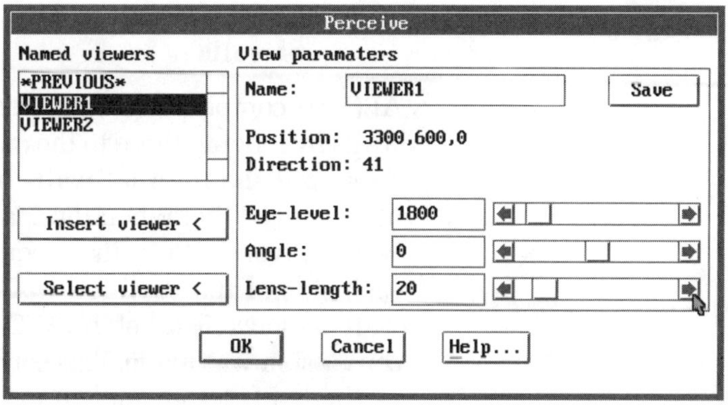

Fig. 2.23

The results of your perception can now be seen on screen. If you are too close the view will be distorted (Fig. 2.22).

The view looks worse in Fig. 2.23 as changing the zoom lens length brings the image even closer.

To improve things go back to the plan view and Zoom, Scale X to 0.5 of full size. Then insert another viewer further back but use the same values for eye level, zoom lens length and view angle. The view now looks much better, as shown in Fig. 2.24.

Note   Note that viewers are inserted at ground level, not at wall base level which is 900 mm below ground (in this case).

7   At this stage further viewers can be placed on screen to create views in any direction, either inside or outside of the building.

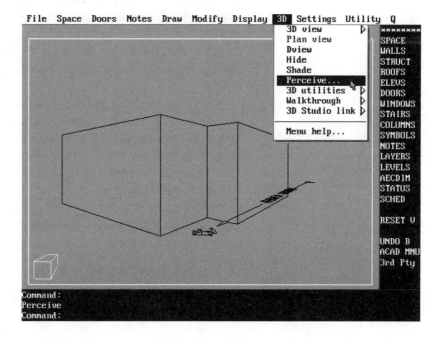

Fig. 2.24

Combining space diagrams with direct placement of walls

AEC 3.0 compensates fully for the space taken by the external walls by moving these to the outside of the space diagrams but can only split the internal walls between adjoining rooms. You will have noticed that when the space diagrams in the previous section were converted to walls an equal amount of space was taken from each side of the partition walls. Half the inner leaf of the cavity wall was also lost before AEC Version 3 when the space to walls conversion was made. This could be avoided by combining the two methods of design and drawing the external walls around the outside of the space diagram which has not been converted to a walls plan.

Set up wall alignment and justification prior to drafting so that no space is taken from the space diagram perimeter. The following section has full details of wall settings.

2.2 Internal partition walls

This area of work is covered in depth in the AEC manual so only a few routines and tips for partition wall creation are covered in this section.

The intermediate walls in the showroom drawings are drawn using Draw wall … from the Walls pull-down menu or the tablet [J1–2]. Typical popular routines are shown in the figures and described below. Solid walls to toilets may need to be smaller than walls previously set and the wall height needs to be set up prior to drawing each wall that differs in height. Wall lines will still show junctions above partitions as though a wall existed above the set height.

Wall heights need to be jotted down on a notepad as they are established or use the AutoCAD List command to find the wall height (Z thickness) for any section at any time. Partition walls with sloping tops can also be drawn using Draw wall … from the Walls pull-down menu or the tablet [M2].

Relative from and to points which mark the beginning and ends of walls can be entered precisely from the keyboard by using the Draw wall … routine and entering data at the command line, such as:

Prompt: To point: **@5678<245** *enter*

This means that the wall will finish 5678 mm from the origin point and incline at an AutoCAD angle of 245 degrees. For building designers, who unlike engineers have the benefit or curse of standardised material sizes, the graphical method of fixing points to a Snap or Grid intersection will usually be quicker and easier. Refer to the detailed routines that follow this introduction for creating various types of wall.

If you need to set up construction lines from which to start internal walls refer to any book on basic AutoCAD.

Temporary marker lines can also be useful for setting positions for the start and finish of walls as can the OSNAP MID command if a single room is to be split exactly in half as shown on the following pages.

Drawing single walls is also useful for internal features such as shop or bar counters which can be brick single walls topped with a simulated flat roof.

Even in later releases of AEC the wall clean up routines are not completely reliable; the last resort, if the clean up procedures fail completely, is to use the trim command to remove individual lines. However the routine is very slow and tedious compared to the automatic AEC routines provided.

Practical

During this session you will practise the following:

- Dividing up internal spaces with single walls.
- Constructing wall elements within the building.
- Cleaning up intersections where walls meet.
- Changing the height of walls after placement.

Methods of placing partition walls

Methods of positioning partition walls so that they are drawn a precise distance from other walls or features are many and varied. Those listed below are a guide only and are far from an exhaustive list of aids to precise positioning.

To draw single partition walls using midpoint OSNAPs to divide an existing room in half.

1   Zoom a window around the area of drawing where you need to work by picking [K9] and defining two diagonal points (top left and bottom right).
2   Pick Draw wall ... from the Walls pull-down menu (Fig. 2.25) or the tablet [J1–2].

The wall dialogue box appears on screen as shown in Fig. 2.26

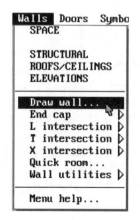

Fig. 2.25

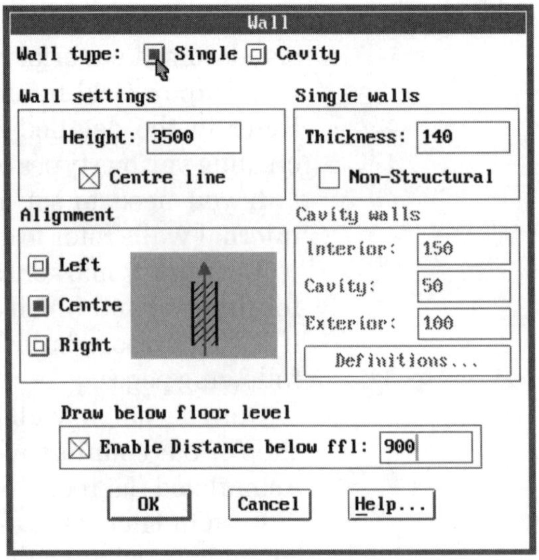

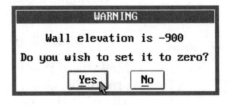

Fig. 2.26

3   Wall type: pick Single.
4   Alignment: pick Centre.
5   Single walls: Thickness: 140 and enable the Non-structural setting.
6   Wall settings: Height: 3500 and enable the centre line setting.
    Comment   Centre lines can be useful for dimensioning later.
7   Click on OK.

A dialogue box will pop up before you can start as shown in Fig. 2.27.

Fig. 2.27

8   Click on Yes.
    Comment   Set the wall lines to 0 for non-structural walls starting at ground level.
9   Prompt: BETween/RELative/<From point>: pick OSNAP MID on the inner wall line as shown in Fig. 2.28 below.
10 Prompt: RELative/<To point>: pick OSNAP MID on the opposite inner wall line as shown in Fig. 2.28.

Before cleaning up the wall intersections turn off the wall centre line layer so that the centre lines are not picked by mistake.

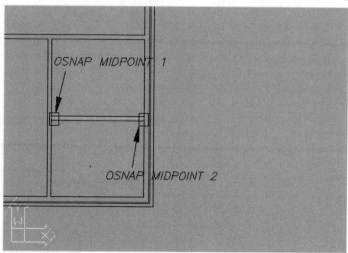

Fig. 2.28

Fig. 2.29

Fig. 2.30

11 Pick Layer states, Off from the Layers pull-down menu or the tablet menu (Fig. 2.29).
12 Select PICK single entity from the Layers pull-down menu.
13 Prompt: Pick entity: select any wall centre line as shown in Fig. 2.30.

All the wall centre lines will now disappear from the screen.

Cleaning up wall intersections

To cleanup any junction between walls.

1 Pick Wall cleanup on tablet [K2] or T intersection from the Walls pull-down menu as shown in Fig. 2.31.
2 Prompt Aperture/Dragbox/Select/<Pick intersection>: **D** *enter*
3 Pick the intersection with dragbox 1 by defining two points as shown in Fig. 2.32.
4 Prompt: Indicate leg of T: pick point as shown in Fig. 2.32

Watch the wall clean up take place on screen. Now repeat the process at the other end of the wall as shown in Fig. 2.32 using drag box 2.

Note If any zoom command has been used previously always use the drag box option and select two points picked either side of the intersection. Whenever both cavity and single walls are involved at intersections always pick the single or single/cavity wall clean up option.

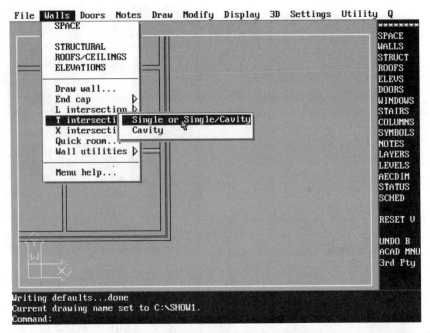

Fig. 2.31

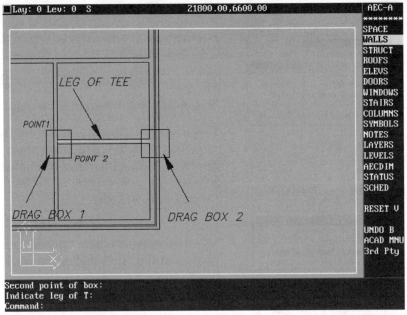

Fig. 2.32

Dividing the toilet area

We are now going to divide the toilet up into cubicles and a lobby. From the space design method we know that the toilet has a width of 2850 mm. Use the AutoCAD distance command if you want to double check.

Fig. 2.33

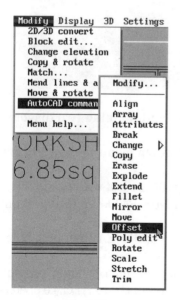

Fig. 2.34

The lobby width has been set at 900 mm.

Using offset guidelines to draw partition walls

Offsetting a cavity wall line a precise distance can act as guide lines for the partition wall to form a lobby (Fig. 2.33).

1   Pick Offset from the tablet [W14] or from the Modify pull-down menu (Fig. 2.34).
2   Prompt: Offset distance or through <Through>: **900** *enter*.
3   Prompt: Select object to offset: pick vertical outer cavity wall line.
4   Prompt: Side to offset: pick any point on the outside of the cavity wall as shown in Fig. 2.33.

A second line should now appear on screen parallel to the line you selected as shown in Fig 2.33.

The internal partition can now be drawn an exact distance from the external wall face after setting up the wall alignment as shown in Fig. 2.35.

To draw the partition wall as shown in Fig. 2.36 use endpoint OSNAPs on the offset wall for the start point, and finish anywhere beyond the showroom partition. Be sure to toggle Ortho on using the F8 key before drawing the wall.

Note   Wall alignment is set to right to preserve the lobby width for future door insertion.

Now clean up the junctions between the walls as described earlier to produce the view shown in Fig. 2.37.

Note   Note that one line has been left in due to the different wall heights involved. Use the AutoCAD trim command if you really want to get rid of it.

Drawing walls relative to other features

Sometimes you may need to draw a single or cavity wall that starts or finishes a known relative distance and angle from another feature. This can be achieved by entering the relative option at the initial prompt or at the final prompt or both.

To draw a small cubicle in the showroom (as a semi-private negotiating area, for instance) use Relative picks to set accurate distances for starting and stopping walls:

1   Zoom a window around the area of drawing where you need to work by picking [K9] on the tablet or picking Zoom window

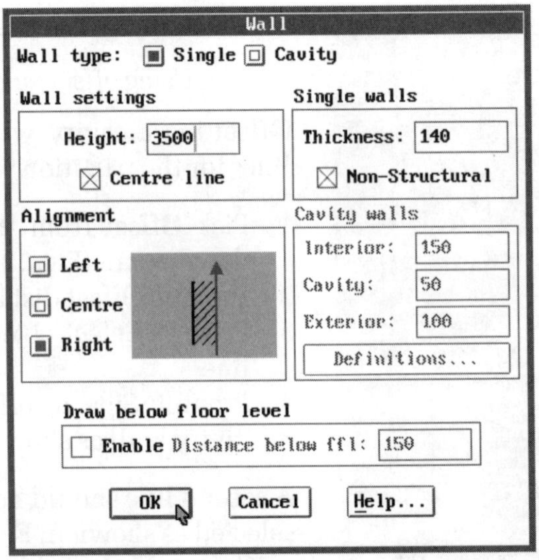

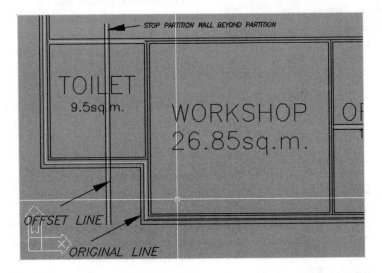

Fig. 2.35

Fig. 2.36

from the display pull-down menu and defining two diagonal points.

2   Pick Wall from the Q (Quick) pull-down menu as shown in Fig. 2.38.

    Comment   Quick Wall has settings of the last wall used.

3   Prompt: Arc/BETween/Length/New/RELative/Undo/<From point>: **REL** *enter.*

4   Prompt: Relative base point: pick OSNAP NEA then pick point on wall as shown in Fig. 2.38.

    Comment   An offset line or temporary marker could provide an exact pick. The Wall offset command [J3] in AEC 4.0 speeds up offset wall drawing.

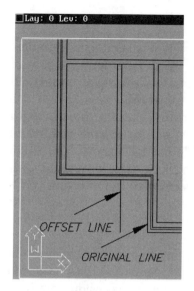

Fig. 2.37

5   Prompt: Displacement/RELative angle <0>: **180** *enter.*
    Comment   See DDvpoint to remind you of the AutoCAD angles convention.

6   Prompt: Relative distance: **2000** *enter.*
    Comment   This sets up the start point for the wall.

7   Prompt: Arc/BETween/Length/New/RELative/Undo <To point>: **@1000<180** *enter.*
    Comment   Walls can be drawn at precise angles and distances.

8   Prompt: Arc/BETween/Length/New/RELative/Undo <To point>: **@1500<90** *enter.*
    Comment   Use the Undo option if you make a mistake on the length or angle.

9   Prompt: Arc/BETween/Length/New/RELative/Undo <To point>: **@2000<0** *enter.*
    Comment   Use the RELative option if you want to finish the partition wall an exact distance from the external wall.

The wall should now look like the one shown in Fig. 2.38

10 After all this you may think that the wall is to thick and want to make a change by typing in a new value. Unfortunately you can only change the height so the wall width will have to be reset and drawn again.

Finally, finish the wall by capping each end using Wall cap [K4] on the tablet, or from the pull-down menu.

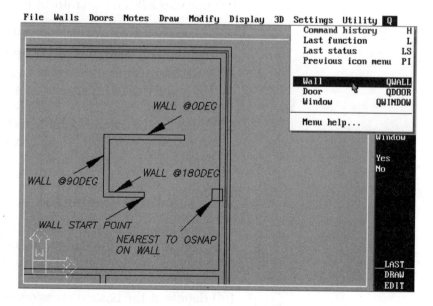

Fig. 2.38

Error messages

One of the major problems that new AEC users have is under-
standing the error messages that can result from faulty setting of
wall parameters. These must be dealt with at this stage or they will
crop up over and over again as new features are added to the
drawing. Each new release of AEC tries to present the user with
better ways of dealing with error messages and intelligent
prompting methods are used to guide users decisions. This often
involves queries about individual walls or even individual wall
lines which require yes/no answers to resolve the insertion problems.

Intersection error messages

A rich variety of error messages can be generated by intersection
cleanup eg walls out of parallel, concurrent wall lines, even too
many arguments has been known to appear on occasion. Some can
be cured by placing a drag box around the wall – others by
answering yes to a prompt – but if the problems persist you can bet
your boots that inserting doors and windows is going to be painful
later so it may pay to start again.

Wall line error messages   If you get an error message concerning
number of wall lines when cleaning up intersections check how
many were included in the pick box. Then try picking just the
inner leaf of the cavity and the single wall. If you are near a
junction check if the wall returns have been included.

Typical messages and solution

- Walls are not parallel: toggle Ortho on before drawing walls.
- Too many arguments: reposition pick box.
- Too many lines, 5 found expects 4: reposition pick box to enclose
  fewer lines.
- 3 walls and 2 cavities found: cavity walls have been redrawn
  over existing wall lines so redraw cavity wall.
- None concurrent wall lines: indicate wall with drag box.

2.3 Dimensioning the model

To the engineering designer/draughtsman accurate dimensioning
is a critical part of the design process and this is reflected in the
ever-increasing diversity of features as each new version of
AutoCAD is released. Many works published on AutoCAD contain
full details of the basic dimensioning procedures for mechanical
and electrical engineers.

For the architects or civil engineers form is just as important as size and tolerances will usually reflect what can be achieved on site rather than theoretical limits. Building designers will remember to use realistic tolerances in accordance with the needs of tradesmen on site when producing working details.

AutoCAD AEC contains many enhancements to dimensioning commands which allow variation of style and form for the Architect and construction designer which may be configured before use or edited later. So many variations that setting up your own personal or company style can be complicated. As the dimensioning information is already contained in the drawing database displaying this information should be much simpler.

Now is a good time to put some initial check dimensions on the plan view; when you start to edit the drawing the dimensions change automatically (associative dimensioning) as the client changes his/her mind.

The dimensioning routines produce dimensions that are specific to AutoCAD and AutoCAD AEC but, unlike some 3D design packages, are 2D only, like text. Dimensions can be placed on a number of separate layers associated with entities so that visibility can be easily controlled.

Some of the options that can be picked are illustrated below and the insertion routines are very simple to use after a little practice even though they are not well structured. If you are a detail drafter working only in 2D there will be little here to help you but detailing all the options and possible situations where dimension variables will be needed would take at least 50 pages. Refer to one of the many texts available on AutoCAD for more detailed information on this very intricate topic.

Stretching and shrinking space and construction elements in horizontal and vertical planes can easily be achieved using AutoCAD AEC. On many building projects even standard designs can be changed without changing the cost of construction significantly. The ability to stretch rooms, walls and even entire buildings is therefore extremely useful. It also enables detailed space analysis to be undertaken on sites where room for future expansion might be limited. Alternative floor plans can also be developed very quickly based on similar schematic layouts but giving very different accommodation.

Note   This is one of the best routines in AEC and a real time saver so practice it on all the design models in each chapter.

Practical

During this session you will practise the following:

- The use of some simple horizontal and vertical dimensioning features to show the sizes of building elements.
- Checking design spaces for suitability using dimensioning features.
- Altering previously drawn elements to provide a better design solution.

Dimensioning the showroom cubicle

As a practical introduction to a very complex topic follow the example listed below and shown in Fig. 2.39 below to put some simple dimensions on the plan view of the showroom cubicle created in the previous section.

1   Pick AECDIM status from the AECDIM pull-down menu or from the template [V3] to set general parameters for the dimension status as shown in Fig. 2.39.
2   Pick the settings as shown in Fig. 2.39 if you want the dimension text shown, otherwise click Reset to defaults.
    Comment You will be automatically presented with this dialogue box only once when you start dimensioning a new project
3   To bypass the never ending cascading menus enable Dialogue mode; you can access all the dimensioning commands from one dialogue box as shown in Fig. 2.40.
    Comment If you prefer to use the pull down menus leave this option blank. The pull down menus and the dialogue box do duplicate one another, yet another cause for confusion to the new user.
4   Pick the vertical option from the dialogue box or from the pull-

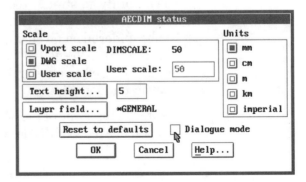

Fig. 2.39

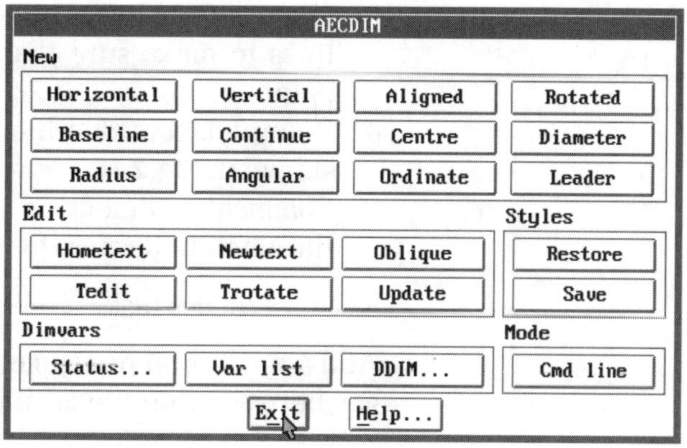

Fig. 2.40

down sub-menu as shown in Fig. 2.41.

5 Prompt: Layer <General>: *enter.*

Comment Dimensions can be associated with a wide range of construction elements or inserted as general entities not related to a particular element. This feature as implemented is a half way house to true elemental dimensioning which many designers would like to see included in the package.

6 Prompt: First extension line origin: pick OSNAP END or *enter* and select first corner.

7 Prompt: Second extension line origin: pick OSNAP END or *enter* and select first corner.

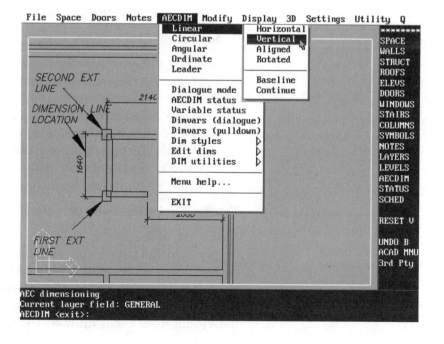

Fig. 2.41

Comment   Always use OSNAPS when asked to pick extension
lines to make sure that the dimension starts from the origin
points.

8   Prompt: Dimension line location: pick any suitable position as
shown in Fig. 2.41.

Comment   Dimension line location can be any distance from
the element it refers to.

Practice task

Add a horizontal dimension to the cubicle side and dimension the
gap between the cubical and the wall.

Changing dimension parameters

If you are not happy with the text height on screen or need to
change the dimensioning units pick AECDIM status again before
continuing with more dimensions.

1   Pick the AECDIM variables menu as shown in Fig. 2.42 to gain
access to all of the varied settings that can be attached to
dimension text, such as arrow size, etc. (see below).

2   Select DIMASO from the menu as shown and make sure that it
is toggled on.

Comment   The stretch command in the following section will
not work unless this is done.

Other useful settings

Some other useful settings from the AECDIM variables list are:

DIMBLK   Arrow type (use Default or pick from icon menu)
DIMASZ   Arrow size (use 7 mm)
DIMALT   To show metric and Imperial units at the same time
DIMTSZ   Tick size: value = size, 0 = draw arrows
DIMTXT   Text size (use 7 mm)
DIMTAD   Text positioning: 1 = above line; 0 = within line
DIMSHO   Show associative dimensioning whilst dragging
         (important for next section)

The list is in two sections; only the first is illustrated in Fig. 2.42.

Note   At this stage in your learning process do not delve to
deeply into the dimension variables section where you can get
bogged down with detail not needed for this project. The dimensions variables and utilities menus contain almost every conceivable option for dimensioning any type of drawing. If each
individual designer is allowed complete freedom to choose any

```
AECDIM  Modify  Display  3D  Settings
   DIMALT    - Alternate units
   DIMALTD   - Alt units decimal places
   DIMALTF   - Alt units scale factor
   DIMAPOST  - Alt units text suffix
   DIMASO    - Associative dimensioning
   DIMASZ    - Arrow size
   DIMBLK    - Arrow block
   DIMBLK1   - Seperate arrow block 1
   DIMBLK2   - Seperate arrow block 2
   DIMSAH    - Seperate arrow blocks
   DIMCEN    - Centre mark size
   DIMDLE    - Dim line extension
   DIMDLI    - Dim line increment
   DIMEXE    - Extension line extension
   DIMEXO    - Extension line offset
   DIMLFAC   - Length scale factor
   DIMPOST   - Dim text suffix
   DIMRND    - Rounding value

                 more
                 EXIT
```

Fig. 2.42

variation the time taken to set up individual styles would be very considerable. A named person within the practice or company should have the responsibility for setting up and maintaining the standard way of dimensioning and preparing a manual for the guidance of all users. To list all the options available on these menus would take many pages but there will be an answer here somewhwere to almost every conceivable dimensioning problem if you have the time to look for it.

Editing spaces and construction elements

A design decision has been made to provide internal and external access to the toilet lobby and space is looking a little tight for external door insertion and hand washing facilities so you can widen it by 300 mm to make more width.

Note When editing RIBACAD drawings (see Chapter 5) the stretch command is particularly useful as a form of dimensional control both for elevational and sectional symbols.

To stretch the toilet lobby

1 Zoom in on the area to be stretched by using Zoom window and picking two corners to enlarge part of the screen.
2 Dimension the toilet and lobby as shown in Fig. 2.43 below using the OSNAP MID to fix the dimension line ends.
3 Zoom out again so that the whole floor plan can be seen.
4 Pick [N2] or [X17] on the tablet or Stretch from the Wall utilities sub-menu of the Walls pull-down menu as shown in Fig. 2.44.
5 Prompt: Select objects Window/Crossing/etc. <Point>: use the

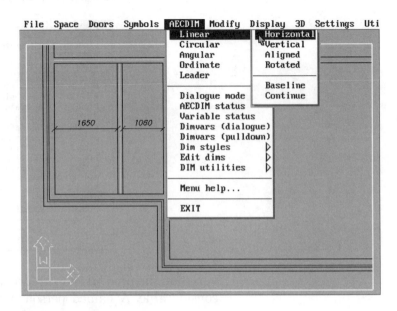

Fig. 2.43

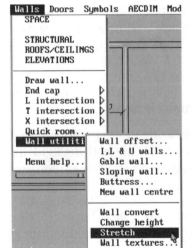

Fig. 2.44

Window option and pick points shown in Fig. 2.45.

6 Prompt: Define base point: pick OSNAP INT on the internal upper left wall intersection.

Comment Pick another point if easier.

7 Prompt: Second point of displacement: type in **@300<180** *enter* to extend the space by 300 mm.

Comment You can use the rubber band to do this if the Snap command is set on. Set Snap at, say, 112.5 mm so that brick coursing can be maintained, otherwise at, say, 100 mm or 300 mm if a dimensionally coordinated grid is in operation.

8 Zoom the view to see the result and note the changed dimension in Fig. 2.46.

Note Make sure that Ortho is toggled on and that the entities picked are only in one horizontal or vertical plane. If not a tangled web of lines will result.

Practice task

Suppose that the client has selected a new standard toilet cubicle depth which is 1850 mm not 1650 mm; try another stretch on your own to accommodate the new toilet module.

Changing wall height

As all the walls have now been drawn for this model this is a good time to try the Change height command which can alter the height

File   Walls   Doors   Notes   Draw

Fig. 2.45

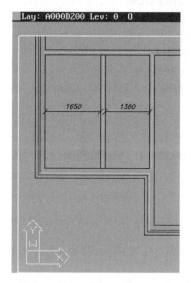

Lay: A000D200 Lev: 0 0

Fig. 2.46

of any pre-drawn non-sloping wall element to a more suitable height should design alterations become necessary.

1   Pick Change height from the Wall utilities sub-menu or the tablet [N1].
2   Prompt: Select walls to edit: select objects: use window crossing or another method to select the required walls.
3   Prompt: Enter the new wall height: **3000** enter or value required.
    Comment   Varying wall heights can be set for different wall elements in the project if required. As suspended ceilings are going to be installed later for each room all walls will be left as the same height for now but could be changed to cut costs later.

        Design check

Now that all the wall elements are in place have a look at a 3D shaded view of the model as shown in Fig. 2.47 before the doors and windows are inserted.

2.4 Automatic door and window insertion

The automatic insertion of doors and windows into walls has been included in AEC from the very first release and was one of the functions that made its reputation with building designers. Today it is still one of the most popular features in the programme, but however well it performs it is basically a feature designed for

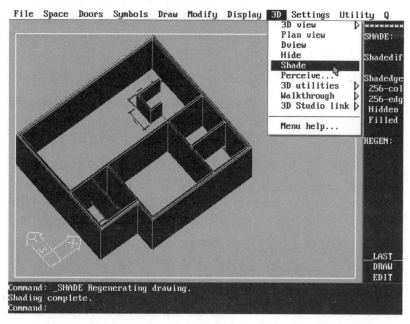

Fig. 2.47

inserting elements into 2D floor plans. In each new release enhancements have improved its performance in 3D at achieving operations such as the insertion of complete window frames. However, it remains an unsatisfactory method of 3D modelling as true holes are not formed in the wall elements – only the faces are pierced. When a window is inserted the wall structure above and below the window is destroyed, and there should be no need to remember to use Wall convert after every tiny modification. This problem would be unimportant if British and continental walls were thin and solid but as they are thick and contain a cavity which is often filled with yet another layer of insulating material this problem is a serious one. This feature of the programme also prevents two levels of window insertion in the same section of wall, which is something that should be possible without setting up two separate wall heights, as should automatic insertion of a gable end windows above the preset wall line. Door and window elements can be inserted at the left hand end or at the midpoint according to the user's choice and the cladding material being used. Where the cladding is sheet or plank material, and distance from the corner is not critical, insertion by the centre point of the window is logical and insertion at a midpoint OSNAP is simple. In western and northern Europe most cladding and some structural materials are heavy, and is therefore made in small units which it is wasteful to cut for the sake of a cheap window frame. Even considering the limitations listed above, in most situations the door and window insertion routines work well and provide a satisfactory basis for 3D visualisation. If you only ever work in 2D then the functions included cannot be bettered for their flexibility and detail.

Practical

In this section you will practise the following:

- Simple domestic door insertion routines in the office and toilet areas.
- Simple domestic window insertion in the office and toilet areas.
- Standard multi-element window insertion features used for the showroom and glazed internal partitions.

Note   AutoCAD AEC users should be aware that the above elements are based on simple rectangular elements and form the starting pint of element insertion. Adding the extensive range of 3D joinery details contained in the programme to these simple insertions can produce complex details that are suitable for working drawings or 3D rendered visualisation. Some transoms and frames have been added to the showroom multi-element window to give an indication to the user of the inserted appearance.

Setting up for insertion

The first stage of any AutoCAD AEC element insertion routine is to set up *all* the status parameters relating to that element. Some changes can be made after insertion but many cannot and much back tracking will be inevitable. It should be possible to combine the status and insertion dialogue boxes for all elements which would remove a source of major irritation for users.

1 Select AEC status from the Settings pull-down menu to access the set up routines, as shown in Fig. 2.48.
2 Open the following dialogue boxes from that menu: 3D Status, Window status, Door status, Wall opening status.
3 Set options as shown in Figs 2.49 and 2.50. Door status should be set in the same way as Window status.

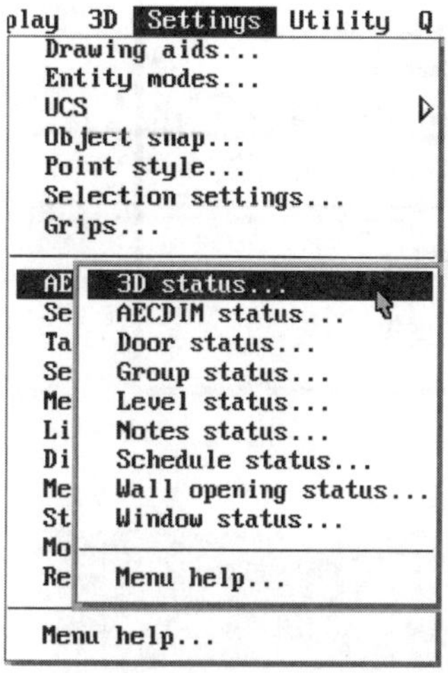

Insertions can be 2D or 3D; all can be changed from 3D to 2D at some later stage, but only some may be changed from 2D to 3D. The door swing is only shown in 2D insertions which are suitable for plan views only.

Choose enable 2D and 3D blocks if both 2D plots and 3D visualisation will be required later. Enable Prompt for symbol elevation so that elements can be set in at any height. Enable Auto viewport update for multiple viewport working later.

See Fig. 2.52 for meanings of window and door status options. See Fig. 2.53 for alternative wall opening status. The Scottish cavity detail has been used in the example insertions as the writer lives in wet and windy hills.

Fig. 2.48

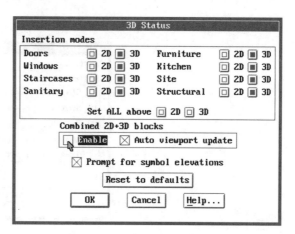

Fig. 2.49

Fig. 2.50

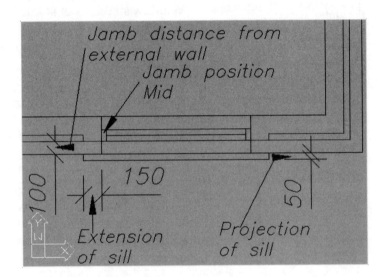

Fig. 2.51

Fig. 2.52

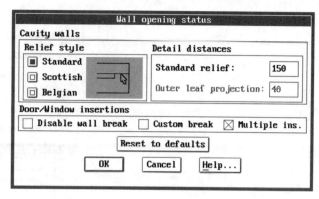

Fig. 2.53

Door insertions

The routines for door and window insertion are very similar but door insertion is made much simpler as the normal insertion point is at the left hand hinge which makes distances from corners much easier to calculate. Midpoint insertion is now possible which works like the mid point window insertion feature.

On splayed walls the rotational angle must be identical to the wall angle. For normal rectangular walls toggle Ortho so that 0, 90, 180 and 270 degrees are the only possible angles for door insertion.

Door frame heads, thresholds and custom jamb details are all supported, as is a library of 3D door types, and ever more joinery details are being included within AutoCAD AEC.

The options for setting Door status are similar to those for windows and meet the demands for more precise settings of insertion and framing parameters.

Simple or complex attributes can be added to inserted doors and windows for automatic 'schedule take off' later.

To insert the external door for the toilet lobby:

1 Pick Door styles then Door leaf styles ... from the Doors pull-down menu as shown in Fig. 2.54.
2 Select option FLUSH_4 from the on-screen list as shown in Fig. 2.55.
3 Pick Door insertions from the Doors pull-down menu (Fig. 2.54) or the tablet [A5–6].
4 Set the dialogue box options as shown in Fig. 2.56 and click on the 3D swing ... box to open it.
5 Set the Swing angle to the required value (Fig. 2.57).

Note   Values range from Closed to 180 (folded back on the wall). If you intend to produce 3D rendered images later set to Closed.

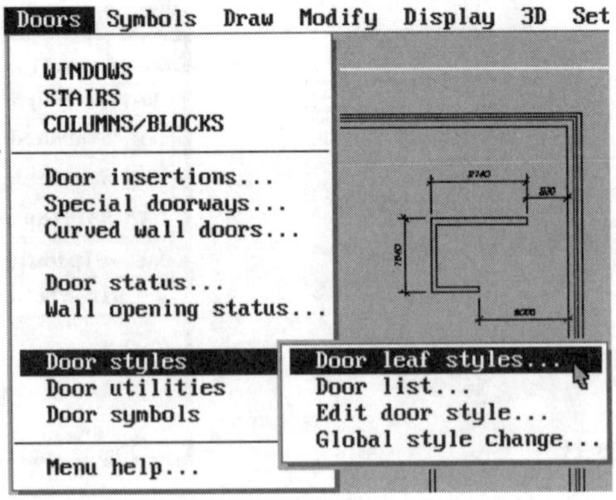

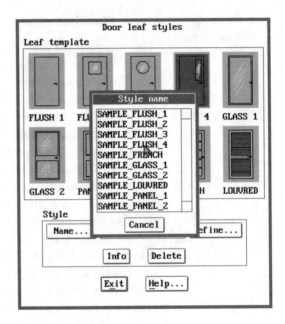

Fig. 2.54

Fig. 2.55

6 Prompt: Symbol insertion point: drag on screen graphically and attach at OSNAP END on partition wall as shown in Fig. 2.58.

7 Prompt: Rotational angle <0>: type in 0 or drag puck with Ortho on to give a visual position.

8 Prompt: Move the symbol? <N>: **Y** enter.

9 Prompt: Second point of placement: drag into a suitable position or use the precise @<distance routine to achieve final placement.

10 Prompt: Repeat door insertion? <Y>: **N** enter.

Comment A multiple door and window insertion feature enables any number of doors or windows to be placed on any one elevation at an equal distance from one another.

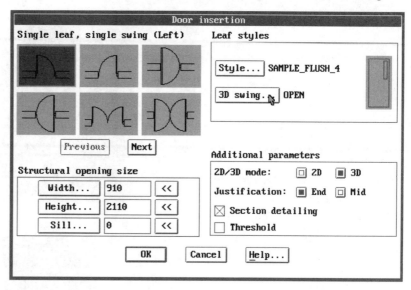

Fig. 2.56

A door should now appear as shown in Figs 2.59 and 2.66 which shows the final placement and door angle.

To convert to 2D plan view of door:

11 Pick 2D/3D convert from the Modify menu or from the tablet [H12–13].
12 Pick 2D mode from the dialogue box shown in Fig. 2.60.
13 Prompt: Select objects: pick any point on the inserted 3D door.
14 The frame and door changes to show the swingline (Fig. 2.61).

Inserting single and multiple windows

To insert single or multiple windows into the same wall for the toilet you first need to divide the area into two equal cubicles and insert standard toilets as shown in Fig. 2.62.

1 Pick Bathroom fittings then Toilet ... from the Symbols pull-down menu or from the tablet [G14–15] as shown in Fig. 2.62.
2 Select Standard toilet.
3 Prompt: Symbol insertion point: using a suitable Snap point attach the toilet assembly as shown in Fig. 2.62.
4 Repeat the insertion or use the Copy command.

Now insert the small high-level windows to the toilet cubicles as a multiple insertion to a vertical wall.

5 Pick Insert window ... from the Windows pull-down menu or from the tablet [A7–8] to insert a window with or without a sill, as shown in Fig. 2.63.

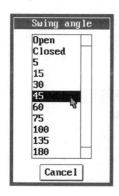

Fig. 2.57

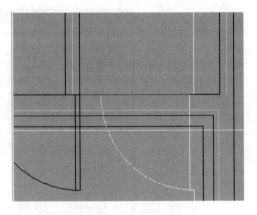

Fig. 2.58

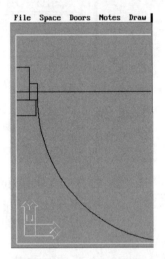

Fig. 2.59

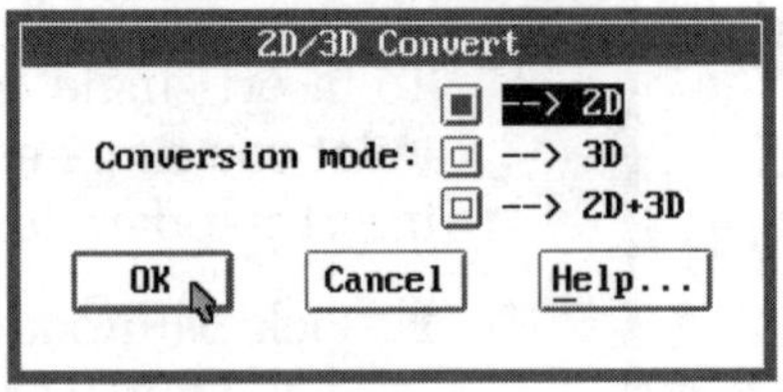

Fig. 2.60

6 Set up the dialogue box as shown in Fig. 2.64 or choose your own settings. Note the dialogue box option meanings in Fig. 2.65.

Note   The Locked dimension (Opening) will be the same for all toilet windows. One dimension – header height, opening height or sill height – can be locked during window insertion.

7 Prompt: Symbol insertion point: use OSNAP to fix on bottom of vertical wall end.

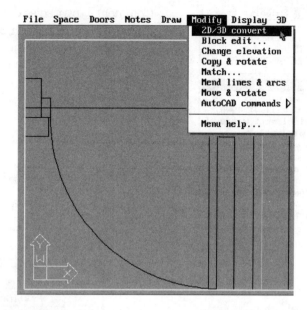

Fig. 2.61

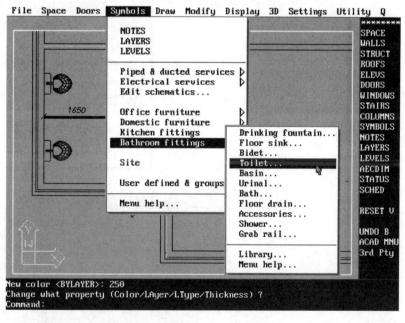

Fig. 2.62

Comment   Symbol can be dragged directly into position but using an OSNAP is safer and more accurate.

8   Prompt: Rotation angle <0>: pick 90 degrees for vertical wall or use the rubber band to drag.

9   Prompt: Move the symbol <no>: **Y** *enter.*

10 Prompt: Second point of placement: drag into position with rubber band or use the precise @<distance routine again.

Fig. 2.63

Comment   A precise distance from the corner can be specified.

11 Prompt: Number of windows to insert along wall <1>: **2** enter.

12 Prompt: Centres/distance between windows: **1066** enter.

Comment   Automatic detection of end or midpoint justification.

13 Prompt: Window array direction <0>: **90** enter.

14 When the required windows have been inserted click on No in the new dialogue box to terminate insertion.

Editing door and window insertions

Several options are available from the Window or Door utilities menus to modify existing door or window insertions. The Adjust

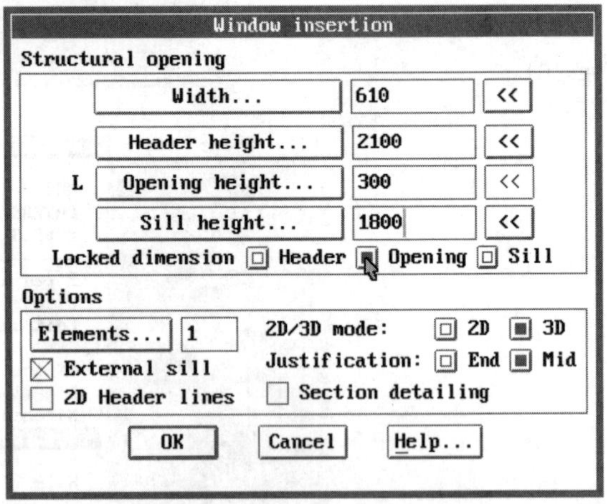

Fig. 2.64

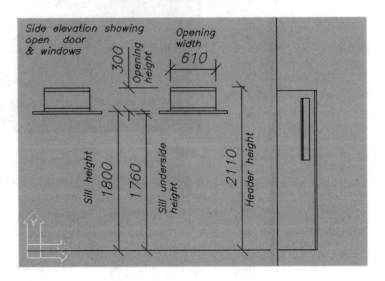

Fig. 2.65

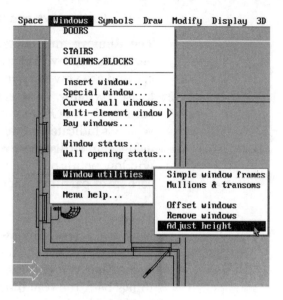

Fig. 2.66

height routine indicates how they can be used.

To adjust the height of an inserted window:

1 Pick Adjust height from the Window utilities pull-down menu
  as shown in Fig. 2.66.
2 Select the required setting from the dialogue box and type in
  the new opening height shown in Fig. 2.67.
3 Prompt: Window the window to edit first point: pick point 1.
4 Prompt: Second point: pick point 2.
  Comment Pick two points that enclose the cavity detail either
  side of the window or some lines will be left unmodified.

Flashing lines on the plan indicate changes to the display in both
2D and 3D.

Note This may be a good time to use the 3D viewing options so
that the changes can be seen and verified from different angles.

Fig. 2.67

Further door and window editing

The Remove and Offset commands were sometimes less than perfect in early versions of AutoCAD AEC but have improved with age and are now very reliable. In both routines you are requested to use a window on screen to select the door or window that needs editing. Under Offset you will be prompted for an offset distance which is indicated either by dragging the symbol along the wall or using the @<distance routine. To ensure success, position the window points outside the cavity closers but away from any wall corners or partition junctions.

Practice

Now try a few window removals and offsets on the windows previously inserted and then try the same with a few doors. Remember you can use Oops, Undo and Redo commands to put things back the way they were. Do not use the Erase command on individual elements in the cavity wall as future editing then becomes impossible. The Reflect swing utility for doors is very useful for enabling rapid insertion followed by equally instant editing when the circulation patterns between rooms are being considered.

Note   Many other features are contained in this section of the programme which are outside the scope of a first course, including production of door and window schedules for contract bills or estimating. Capabilities to produce special door and window openings with non-standard frames are present in the software, but again are outside the scope of a first course.

Shop-front or showroom windows

Now insert the basic multi-element windows which can have 3D elements such as inserted frames, heads, sills, mullions and transoms added later if required. Although this introductory section will only cover the elementary use of multi-element windows they can be used in many situations where standard windows cannot be inserted, for example: high-level glazing, rooflights and dormers. Just picking a start and finish point for the glazing at any level and specifying the opening height is enough for successful insertion. Equal multi-element windows are used throughout this section but if you need the glazing panels to be different widths then pick the Unequally spaced glazing ... option on the pull-down menu and reply to the prompts for further information.

Inserting multi-element windows is a three stage process:

- Freehand sketching on paper of the window layout showing the position and shape of mullions and width of glazing panels (equal or unequal distances between mullions).
- Forming an opening for the glazing by breaking the wall in the correct positions. (Insertion of multi-element windows does not break the wall like other types of window insertion.)
- Insertion of frame and glazing into the pre-formed opening.

To insert multi-element windows into the showroom first break the wall:

1 Pick Door status and set jamb size to 1 mm × 1 mm
Comment This fools AEC into thinking that the jambs are not there and shows an opening without framing which is what you need.
2 Pick Door insertion ... from the Door pull-down or from the tablet [A5–6] and pick the Cased opening option shown in Fig. 2.68 from the on-screen menu.
3 Type in the width, height and sill height or click on the chevrons to enter dimensions on the drawing as shown in Fig. 2.69.
4 Click OK as usual to confirm choices and exit.
5 Prompt: Symbol insertion point: use OSNAP or freehand drag to position on the wall as shown in Fig. 2.69.
6 Prompt: Rotation angle <0>: *enter.*
Comment Rotation angle depends on the insertion point as shown in Fig. 2.71.

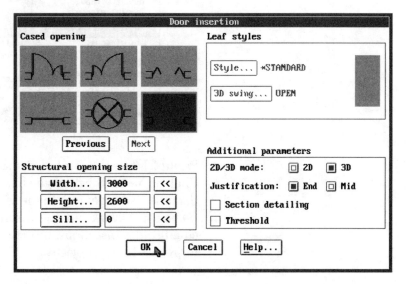

Fig. 2.68

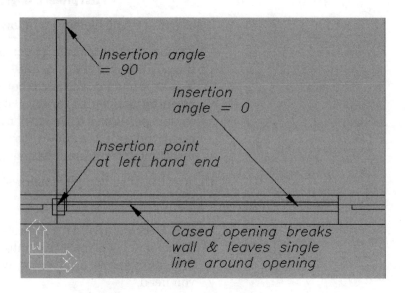

Fig. 2.69

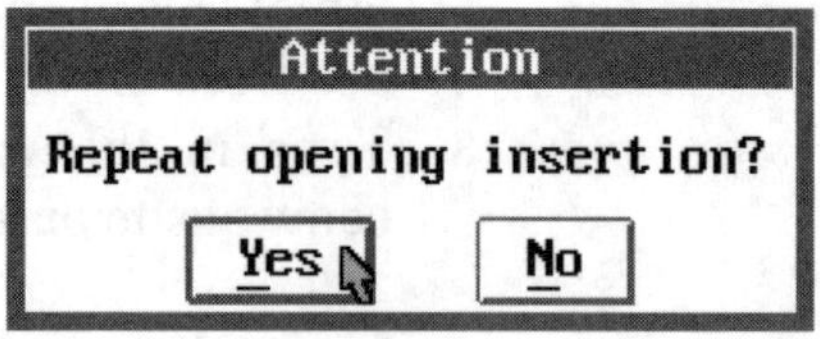

Fig. 2.70

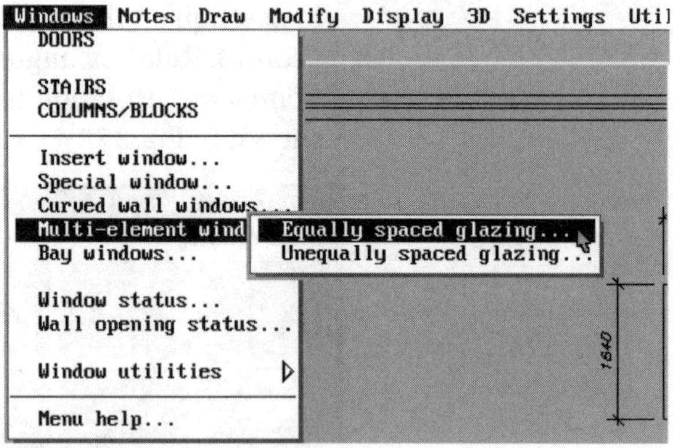

Fig. 2.71

7 Prompt: Move the symbol <No>: *enter.*
8 Click on Yes in the dialogue box as shown in Fig. 2.70 and insert
   the other two openings in the same wall.
   Comment   Work out the distances between the windows to
   equalise the insertions. Inserted elements will not be grouped
   for re-use in the projects.

Multi-element window	
2D/3D Mode:	☐ 2D ■ 3D
Alignment:	Centre ▼
Number of elements:	9
Mullion type:	■ Generic ☐ Custom
Mullion width:	50 <<
Mullion depth:	150 <<
Sill height:	200 <<
Header height:	200 <<
Total height:	2600 · <<
Glazing location:	Middle ▼
Group window:	☐ Yes ■ No
OK Cancel Help...	

Fig. 2.72

You may wonder why the Insert block.. option on the Columns pull-down menu has not been used to form the opening. As with most choices in AutoCAD AEC the reason is ease of editing when changes are necessary and for reliable wall breaking.

Next insert the equal or unequal multi-element windows into the pre-formed wall openings.

9 Pick Multi-element window, Equally spaced glazing ... from the Window pull-down menu or the tablet [C7] as shown in Fig. 2.71.

10 Select the options on the dialogue box that you require, as shown in Fig. 2.72, then click OK to exit.

Note Figs 2.73 and 2.74 show the meanings of the various options on elevation and plan.

11 Prompt: Enter starting point: use OSNAP MID to fix the start point within the wall opening as shown in Fig. 2.74.
   Comment The possible options on this menu produce such a wide range of variations that different architectural styles are produced with each setting. Experiment until the desired effect is produced. Do not forget that the alignment option in Fig. 2.74 must be set left or right before the OSNAP END is used for insertion.

12 Prompt: Ending point: use OSNAP MID on the other end of wall opening as shown in Fig. 2.74.
   Comment The OSNAP setting must be the same for both ends of the wall otherwise a skewed window will result.

The window will now array its various elements on screen in the correct position. Use the Undo command to try again if necessary.

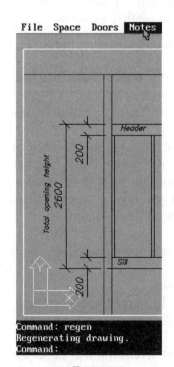

Fig. 2.73

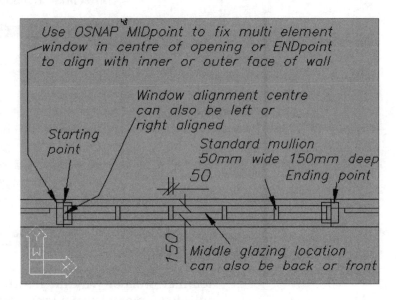

Fig. 2.74

Inserting columns

Rectangular or circular columns are often used to give depth to otherwise plain elevations or to change the perceived relationship between length × height. Structural steel or rainwater downpipes may also need to be concealed in columns added to elevations.

To insert columns for either practical or aesthetic reasons:

1   Pick: Insert column … from the Columns pull-down menu (Fig. 2.75) or from the tablet [A9–10].
2   Pick the options that you require from the dialogue box and click OK as shown in Fig. 2.76.
    Comment   Stick to Mid for the Justification option in the dialogue box.
3   Prompt: Insertion point: use OSNAP MID on the opening or pick your point using any of the methods previously described.
    Comment   Columns will only break one wall if inserted at corners or junctions. The other wall will need a manual clean up operation using the trim and other editing commands.
4   Prompt: Rotation angle <0>: enter.
    Comment   The column Rotation angle must match the wall angle for correct wall breaking.
5   Prompt: Move the symbol <No>: **Y** enter.
6   Propmt: Second point: **@112<180** enter to fix column in position.

    Practice

Insert equal and unequal multi-element windows into the other openings along the same wall then provide a glazed upper parti-

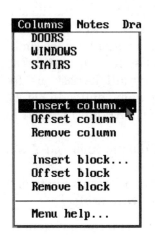

Fig. 2.75

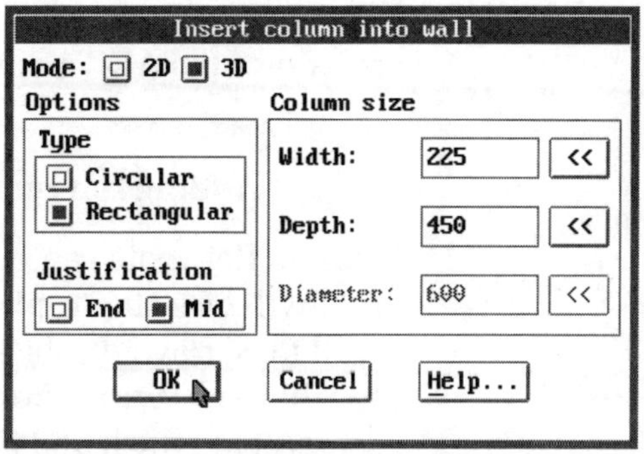

```
┌─────────────────────────────────────────────┐
│           Insert column into wall           │
│ Mode: ☐ 2D ■ 3D                             │
│ Options             Column size             │
│ ┌─────────────┐                             │
│ │ Type        │    Width:   ┌─────┐  ┌──┐   │
│ │ ☐ Circular  │             │ 225 │  │<<│   │
│ │ ■ Rectangular│            └─────┘  └──┘   │
│ │             │    Depth:   ┌─────┐  ┌──┐   │
│ │ Justification│            │ 450 │  │<<│   │
│ │ ☐ End ■ Mid │             └─────┘  └──┘   │
│ └─────────────┘    Diameter:┌─────┐  ┌──┐   │
│                             │ 600 │  │<<│   │
│                             └─────┘  └──┘   │
│        ┌──────┐  ┌────────┐  ┌───────┐      │
│        │  OK  │  │ Cancel │  │Help...│      │
│        └──────┘  └────────┘  └───────┘      │
└─────────────────────────────────────────────┘
```

Fig. 2.76

tion to the sales cubicle in the showroom. View the results in 3D using the perceive command and hide or shade to check out the results of your work.

2.5 Adding roof elements and layer control

You have probably noticed that the design model is becoming more complex as features are added and that the building elements inserted have different colours. This is due to each entity being on a different layer (overlay) that can be manipulated in many different ways and accords fully with BS1192 and other European Architectural design standards. Before eaves details for the roof can be added to the model, you will need to isolate only the layers needed to form the anchor points for the eaves detail. This will then freeze (make invisible and lock) all the layers not needed to construct the roofs in order to reduce congestion of the screen and enable accurate picks to be made. It is fortunate that layers can be isolated by picking single items or elements grouped as entities on screen and full details are included on Layer isolate and Layer off functions in the practice section. The new British Standard layering conversion that has now replaced the self explanatory system first adopted by Autodesk is very complex and best avoided at introductory level if possible. It may conform to all the best principles of library data storage but many practising designers use their own layering system and regard the British Standard as a mish-mash of alphanumerical stupidity that only a committee could invent. On later projects other layers will need manipulation through the AEC complex dialogue box to cut down on regeneration time and speed up the drawing process and this process wil be covered in later chapters.

Later versions of AEC reintroduce many of the self-explanatory layer names from the original version as Layer aliases so that the user can recognise them at a glance and not have to continually reference the AEC manual.

Adding flat roof elements

Simple flat roofs can be constructed as a single thickness but the roof will be a single-plane 3D face which must be raised above the wall tops to avoid coincident vertical wall faces. It is constructed to finish flush with the outer walls and the fascia will also be constructed flush and then stretched to provide an overhang.

The best method of flat roof construction uses the complex floor option and involves setting up an overhanging fascia board that overlaps the wall top for realism and allows scope for the addition of firring pieces, tilting fillets, upstands and downstands if required. This method is more detailed than the one described here and would take up far more space than I have been allowed in this introductory volume.

Practical

During this session you will practise the following:

- Construction of a simple flush flat roof.
- The use of simple layer commands to help construct the model.
- Construction of a fascia and the stretch of the roof assembly to form an overhang.

Creating the roof

1   Pick Roof constructions ... from the Roofs pull-down menu or the tablet [A3–4] as shown in Fig. 2.77.
2   Pick the Flat roof option from the dialogue box options and click OK as shown in Fig. 2.78.
3   Type your required roof level in Height above level box and click on OK (Fig. 2.79).
    Comment   Set the roof level 20–25 mm above the wall height to prevent wall faces showing through the roof plane when using the Hide command.
4   Click on Yes in the OSNAP END dialogue box as shown in Fig. 2.80.
5   Prompt: Draw a continuous line indicating the perimeter box from point: pick a corner of the roof.
6   Prompt: Arc/BETween/Close/Length/RELative/Undo/   <To

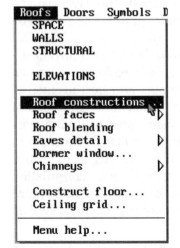

Fig. 2.77

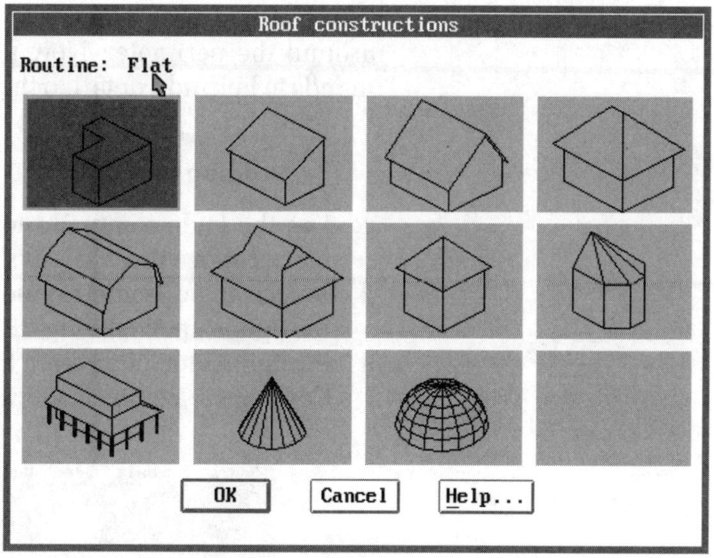

Fig. 2.78

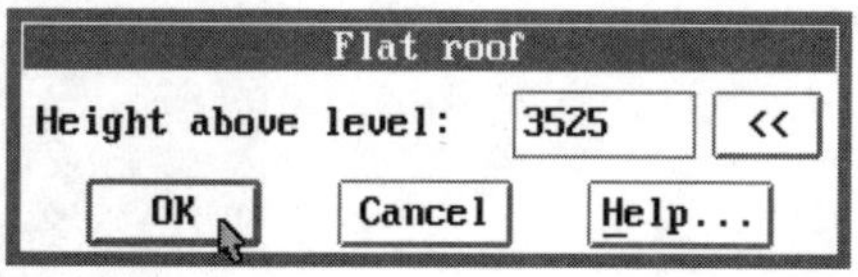

Fig. 2.79

point>: pick the next corner in either a clockwise or anticlockwise direction.

7 Prompt: Arc/BETween/Close/Length/RELative/Undo/ <To point>: continue to pick corners around the roof travelling in the same direction.
Comment Use the Zoom window command to enlarge small areas for easy corner picking and Zoom previous to see the full roof again. Do not use Zoom at all during the pick sequence.

8 Prompt: Arc/BETween/Close/Length/RELative/Undo/ <To point>: type **C** *enter* when one point away from the starting point.

9 Prompt: Arc/BETween/Close/Length/RELative/Undo/ <To point>: *enter* to finish sequence.

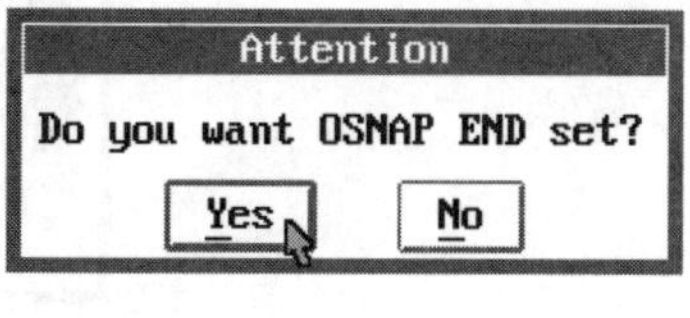

Fig. 2.80

The finished roof plane will appear on screen as a single line around the perimeter. Now use simple layer manipulation commands to help add detail to the roof.

Adding roof detail

1 Use the Hide command to check that a full 3D face has been placed above the walls as shown in Fig. 2.81.
2 Switch to an isometric view to add the roof detail by picking the option shown in Fig. 2.82.
3 Prompt: Enter angle from the XY plane <90>: **+30** enter.
    Comment   Isometric views are really useful for 3D insertions.

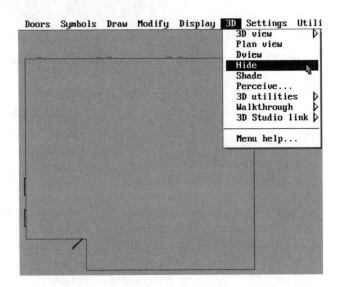

Fig. 2.81

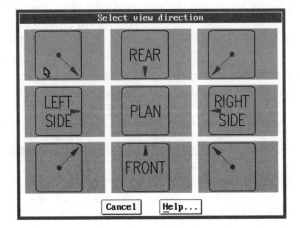

Fig. 2.82

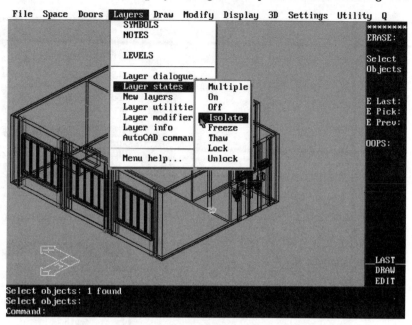

Fig. 2.83

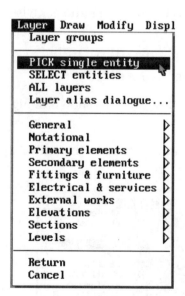

Fig. 2.84

4   Pick Isolate from the Layer states option on the Layers pull-
down menu or from the tablet [Q1] as shown in Fig. 2.82.
Comment   Its much easier to pick all layer commands from the
tablet due to the number of pull-down menus needed to locate
specific options.

5   Pick Single entity as shown in Fig. 2.84
Comment   If you do not feel confident in picking the right
layer from the jumble of lines on screen practice on the
dimension lines inserted earlier. Just pick Layer off from the
tablet [P2] or from the Layer status menu and pick any point on
the dimension lines to banish them from the screen.

6   Prompt: Pick entity: select any point on the roof plane perimeter and the screen should look like that shown in Fig. 2.85.

Adding a fascia

7   Pick [C4] from the tablet or Eaves detail, Fascia board ... from
the Roofs pull-down menu as shown in Fig. 2.85.

8   Enter values in the dialogue box to suit your roof and click on
OK as shown in Fig. 2.86.
Comment   Fascia alignment can be right, left or centre. The
fascia elevation is set to provide a 25 mm upstand above the roof
plane level.

9   Prompt: Arc/BETween/Close/Length/RELative/Undo/   <To
point>: pick each corner point on the roof in sequence travelling
in a clockwise direction.

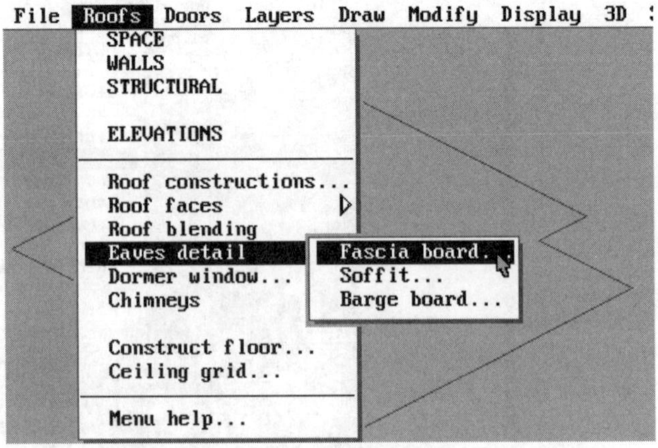

Fig. 2.85

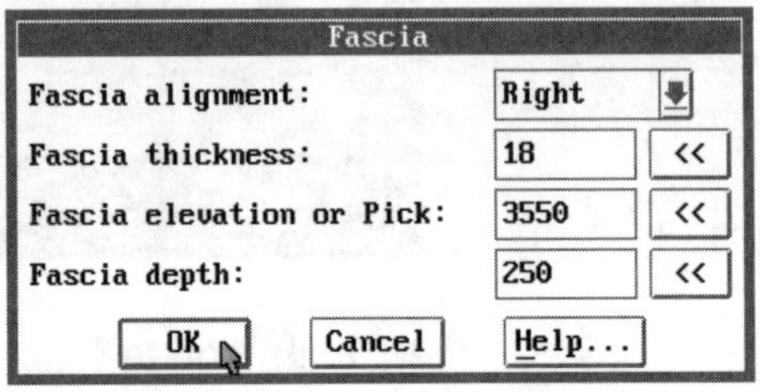

Fig. 2.86

Comment   If you prefer to travel round the roof anticlockwise the fascia alignment should be set left.

The message generating mesh appears to show that roof is now a surface mesh which can be manipulated as one entity.

10 Pick Hide to check that the appearance is OK, as shown in Fig. 2.87.

Finally enlarge the mesh to provide an overhang to the walls, if required.

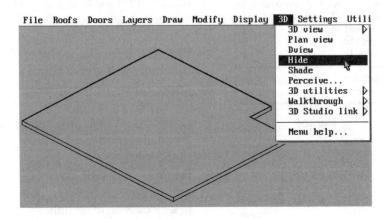

Fig. 2.87

11 Pick Scale from the Modify pull-down menu or from the tablet [W12].
    Comment   This method will not give an exactly equal over-hang on all sides of the roof; Offset the wall lines to form anchor points for the roof if greater precision is required.
12 Prompt: Select objects: window the roof plane.
13 Prompt: Base point: pick a point roughly in the centre of the roof area.
    Comment   Enlarge the roof equally on all sides.
14 Prompt: Scale factor: **1.03** *enter* to change the roof mesh form on screen to show the overhang.
    Comment   Scale factor is calculated from total roof frontage of 12,400 mm and desired overhang of 200 mm at each side of the walls: $12,800 \div [12,800 - (2 \times 200)] = 12,800 \div 12,400 = 1.032$.

2.6 Suspended ceilings and boarded floors

Realistic internal views of a tiled or boarded finish to a solid floor and strip or tiled false ceilings can be easily created within AutoCAD AEC. Visualisation is aided when generating internal views and when standard libraries of service details are available in 3D, space planning of crowded services ducts within floor and ceiling voids will be facilitated using these 3D techniques.

  3D symbol insertions will be used to complete the interior of the showroom and as a basis for further enhancement using rendering techniques provided by AutoCAD or third party developers.

Practical

In this section you will practise the following:

- Insertion of a solid floor to the showroom.
- Placement of a grid pattern on its upper surface.
- Inserting a ceiling grid 500 mm below the roof plane.
- Inserting 3D symbols to complete the showroom model.

Inserting a solid floor

1 Pick Solid floor from the Structural pull-down menu or the tablet [F7].
2 Enter parameters in the solid floor dialogue box as shown in Fig. 2.88 (0 for the elevation; 200 for thickness) and click OK.
3 Using OSNAPS and the Zoom window, Zoom previous commands pick points to define the floor's perimeter.

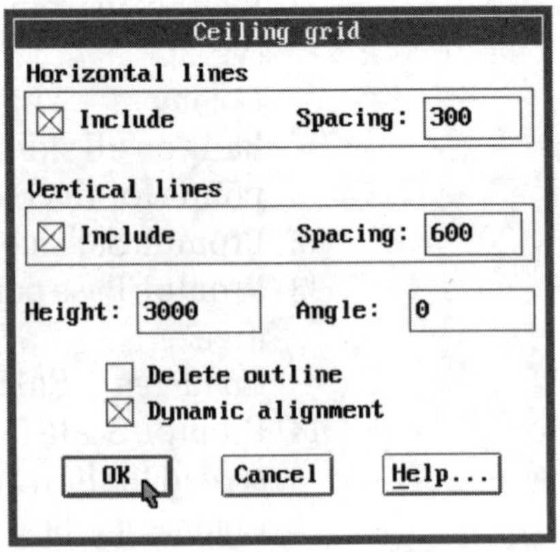

```
┌─────────────────────────────────────────┐
│              Ceiling grid                │
│ Horizontal lines                         │
│ ┌──────────────────────────────────────┐ │
│ │ ⊠ Include        Spacing: │300│       │ │
│ └──────────────────────────────────────┘ │
│ Vertical lines                           │
│ ┌──────────────────────────────────────┐ │
│ │ ⊠ Include        Spacing: │600│       │ │
│ └──────────────────────────────────────┘ │
│ Height: │3000│       Angle: │0│          │
│              ☐ Delete outline            │
│              ⊠ Dynamic alignment         │
│       │ OK │   │ Cancel │  │ Help... │   │
└─────────────────────────────────────────┘
```

Fig. 2.88

4   Use any 3D viewing method to check your results.

Now use similar techniques to place the grids in position.

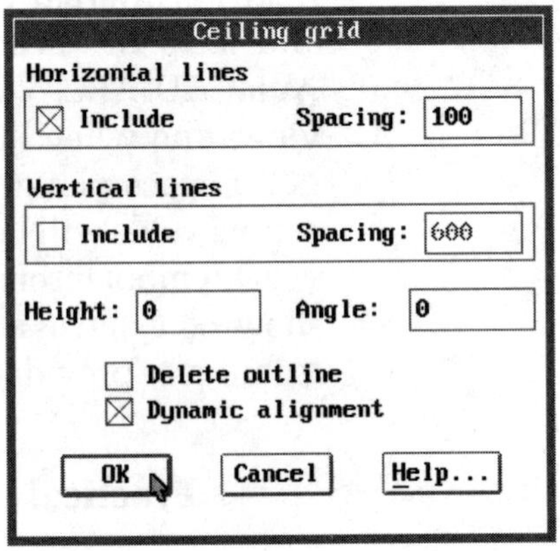

```
┌─────────────────────────────────────────┐
│              Ceiling grid                │
│ Horizontal lines                         │
│ ┌──────────────────────────────────────┐ │
│ │ ⊠ Include        Spacing: │100│       │ │
│ └──────────────────────────────────────┘ │
│ Vertical lines                           │
│ ┌──────────────────────────────────────┐ │
│ │ ☐ Include        Spacing: │600│       │ │
│ └──────────────────────────────────────┘ │
│ Height: │0│          Angle: │0│          │
│              ☐ Delete outline            │
│              ⊠ Dynamic alignment         │
│       │ OK │   │ Cancel │  │ Help... │   │
└─────────────────────────────────────────┘
```

Fig. 2.89

Placing a grid pattern

5   Pick Ceiling grid from Roofs pull-down menu or from the tablet [E3].
6   Enter parameters for the grids in the dialogue box as shown in Figs 2.88 and 2.89 and click OK.
7   Using OSNAPS and the Zoom window, Zoom previous commands pick points to define the ceiling grids perimeter.
8   Use any 3D viewing method to check your results, which should look something like Fig. 2.90.

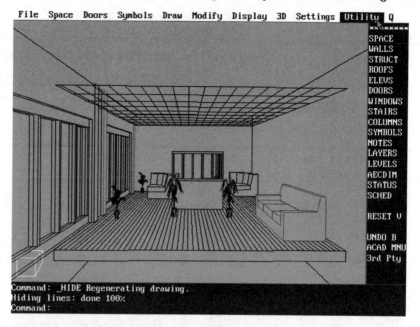

Fig. 2.90

Fig. 2.91

Adding 3D symbols

9 Pick Symbols insertion from the Symbol pull-down menu or from the tablet [D14–15 through H14–15] to complete the interior as shown in Fig. 2.91.

Note The 3D interior views can be rendered or enhanced using specialist graphics packages for visualisation purposes.

CHAPTER 3

Second project: single-storey domestic building

This design project is typical of the current trend towards detached dwelling units for the small family or retired couple. The project uses the direct wall method to design and detail a two/three-bed detached or semi-detached bungalow with detached single garage. Individual dwellings may be designed using this method whereas most estate type dwellings are more economically designed using the space design method. As usual the design process will involve both two- and three-dimensional work allowing hardcopy for working drawings in 2D or 3D images for visualisation purposes. Traditional methods of construction are assumed with a brick/block cavity wall and pitched roof. Window and door insertions described include special door openings and two types of bay window; not those already covered in the first project. Roof design will be confined to the placement of standard pitched roof elements and simple blending of elements.

3.1 Direct wall placement

The method of designing walls described here is most suitable for individual new projects, small new works and renovation/alteration projects.

Practical

During this session you will practise the following:

- Setting up the drawing aids needed to control the design dimensions.
- Defining the cavity wall dimensions to your own specification.
- Drawing the external walls to the bungalow.
- The use the quick room command to add a detached garage.

Bungalow design model

1 Pick Start NEW drawing from the AEC dialogue box.
2 Type BUNG1 at prompt for drawing name or use any drawing name of your choice as shown in Fig. 3.1.
  Comment No spaces are allowed in the drawing name. Space may be used in the longer names allowed by new operating systems.
3 Pick *UNNAMED from the style box and confirm 1:50 as the hardcopy scale.
4 Pick A3 by clicking on the A3 button to black out the centre.
5 Pick the Landscape orientation button.
6 Set plotter margins by typing in new values.
7 Click OK after checking settings.

Fig. 3.1

Next set up the dimension control method:

8 Pick Drawing aids ... from the Settings pull-down menu as shown in Fig. 3.2.
9 Set up Modes, Snap and Grid options in the dialogue box (Fig. 3.3).

Note Grid and Snap values are set up for designing to internal dimensions on a 300 mm grid. If you wish to design to external dimensions then standard sizes of cladding materials are easily catered for. Brickwork could use a 225 mm Grid, with Snap set to 112.5 mm. Refer to the function key toggle given information in Fig. 1.15 (page 000).

10 Pick [J1–2] from the tablet or Draw wall ... from the Walls pull-down menu, as shown in Fig. 3.4.
11 Set the options in the Draw walls dialogue box as shown in Fig. 3.5 and click on Definitions ... to call up the Cavity wall definitions dialogue box.

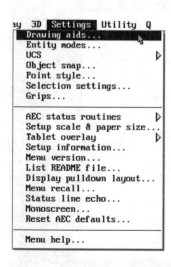

Fig. 3.2

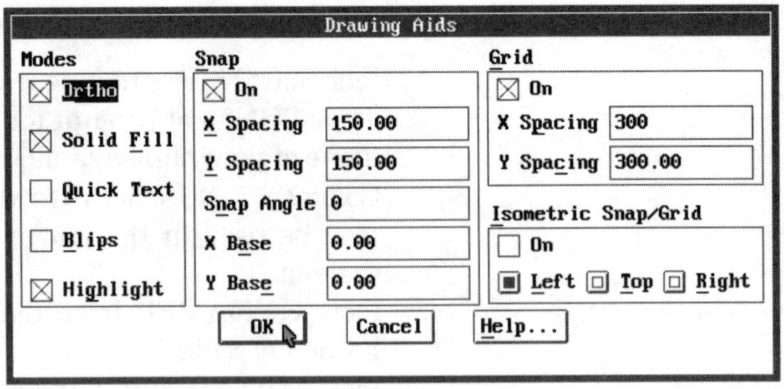

Fig. 3.3

Note   Wall height is set to a standard domestic value, set your own if required. Alignment of the wall is set to right as you are drawing clockwise to internal dimensions. Align to the left if you have set up the grid and snap values for brickwork external dimensions and are drawing clockwise. Note also that the Grid dots are visible on screen and that the coordinate values now reflect the Grid settings.

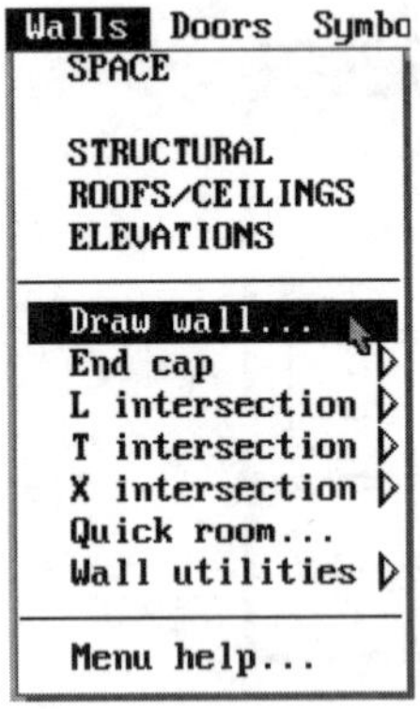

Click on OK if the cavity wall is correctly defined.

If the site is flat enable the draw below floor level option:

12 Type a name in the name box (e.g. SHEP1) and click on Define in the dialogue box as shown in Fig. 3.6.
13 Type values in the Define new definition dialogue box for the new cavity wall, as shown in Fig. 3.7, and click OK.

Fig. 3.4

Fig. 3.5

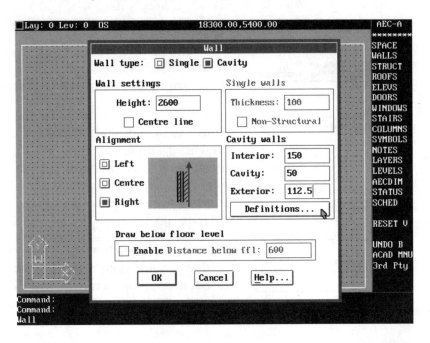

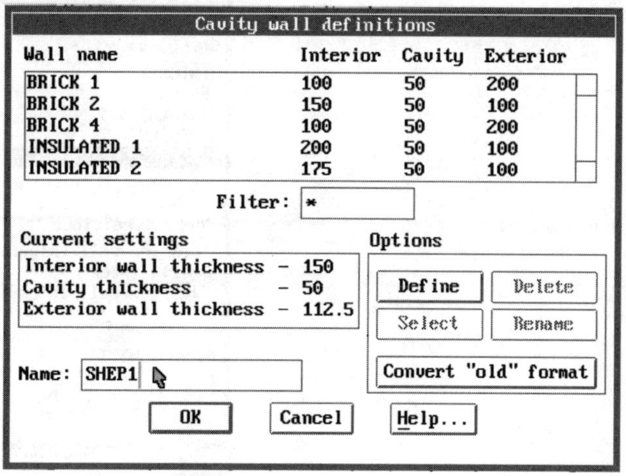

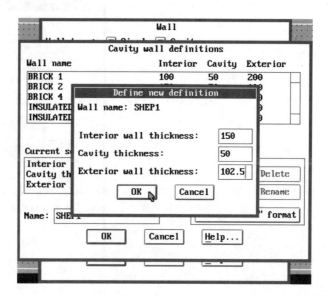

Fig. 3.6

Fig. 3.7

14 Click on SHEP1 which has now appeared in the dialogue box and then on the Select option (Fig. 3.8) and click on OK.
15 Click OK to accept the new values from the Wall dialogue box as shown in Fig. 3.9.

Now the external wall can be positioned on screen, aligned to the internal dimensions through movement of the mouse or puck. Make sure that Ortho is toggled on. Remember that if you are

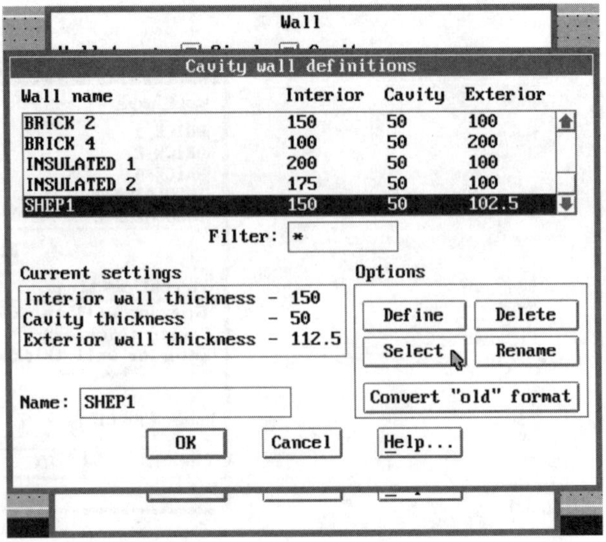

Fig. 3.8

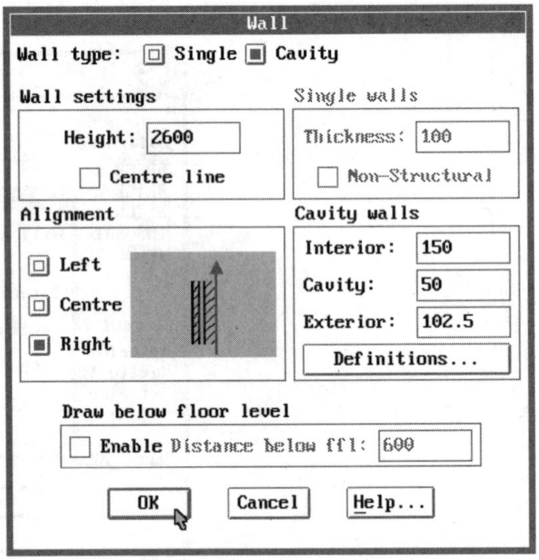

Fig. 3.9

designing to suit the external dimensions, Grid, Snap and alignment settings will need setting to different values.

16 Prompt: BETween/RELative/From point: pick a point near the top left corner of screen.
    Comments   Note how the X and Y coordinates on the status bar change to 0,0 when the point is picked. Make sure that OS is visible on the status bar so that you know that Ortho and Snap are both toggled on.

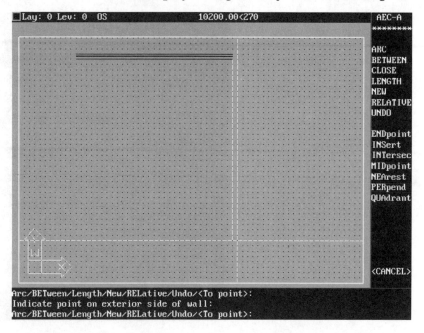

Fig. 3.10

17 Prompt: ARC/BETween/Length/New/RELative/Undo   <To
    point>: drag the cross hairs to the right by 9000 mm (on
    drawing) and pick the point.
    Comments Note how the X coordinate changes at 150 mm
    intervals as you move the cross hairs.
18 The length of wall drawn will now appear on screen as shown
    in Fig. 3.10.
19 Prompt: Indicate point on exterior side of wall: pick a point as
    shown in Fig. 3.11.
20 Prompt: ARC/BETween/Length/New/RELative/Undo   <To
    point>: drag the cross hairs 10,200 mm downward and pick the
    point.
    Comment   If you make a mistake just type **U** *enter* (the Undo
    command) at the prompt to go back one stage.
21 Continue drawing the walls until you reach the bottom left
    corner.
22 Prompt: ARC/BETween/Length/New/RELative/Undo   <To
    point>: **C** *enter.*

The wall will now close back to the starting point and the corner
will be automatically cleaned up on screen. Toggle Grid off to view
the walls.

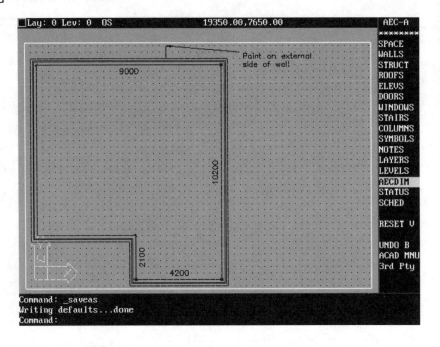

Fig. 3.11

Adding a detached garage

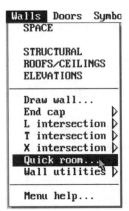

Fig. 3.12

1   Pick [M1] on the tablet or Quick room ... from the Walls pull-down menu as shown in Fig. 3.12.
2   Select the options in the Quick room dialogue box as shown in Fig. 3.13.
3   Prompt: First corner: pick a point on the right-hand side and about 1200 mm away from the side wall as shown in Fig. 3.14.

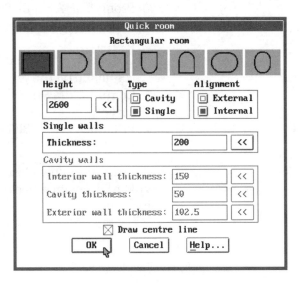

Fig. 3.13

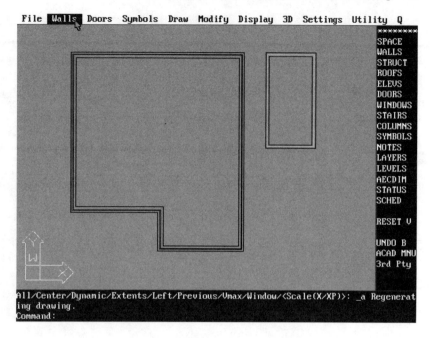

Fig. 3.14

4   Prompt: Dimension/second corner <D>: *enter.*
     Comment   With Snap toggled on (toggle with the function key
     F9) just drag the dimensions to the second corner.
5   Prompt: Horizontal (width) dimension: **2400** *enter.*
6   Prompt: Vertical (depth) dimension: **4800** *enter.*
     Comment   Note that both these dimensions must be non-zero
     and positive.

A single garage will appear on screen as shown in Fig. 3.14. If not
pick the Move command, window the entity and reposition as
above. Pick an isometric viewpoint and generate a hidden 3D
image just to check the wall elements as shown in Fig. 3.15.

3.2 Internal partition walls to the bungalow

You can form these in the same way as those for the showroom and
garage model or use the 'I, L & U' wall method to divide internal
spaces. Although the bungalow project will use this method it will
be on hotel, hospital and office projects that the greatest time
saving will result.

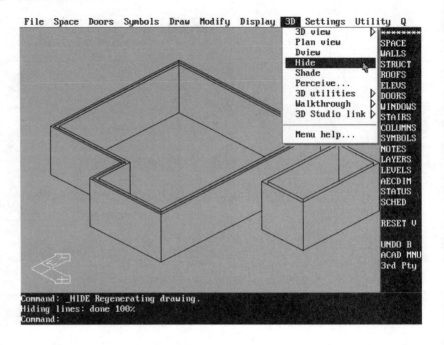

Fig. 3.15

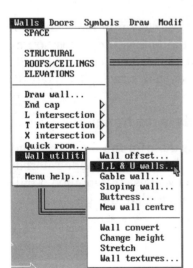

Fig. 3.16

Practical

During this session you will practise the following:

- Dividing the bungalow into internal spaces using the internal partition commands.

Forming the lounge

1 Pick [M4] on the tablet or choose I, L & U walls... from the Walls utilities option on the Walls pull-down menu as shown in Fig. 3.16.

2 Pick the L wall option from the dialogue box and set up the other options to your own choice (Fig. 3.17).

Note Toggle Centre line on if you wish to dimension using CL values. Toggle Intersection cleanup on if using the I wall insertion between internal partition walls, otherwise only the external walls will be recognised.

3 Prompt: Pick space corner (pt 1) <int>: pick top left corner as shown in Fig. 3.17.

4 Prompt: BETween/RELative/Pick point along base wall (pt2) <nea>: **@4900<270** enter.

Comment You could also drag the cross hairs to this point if Snap were toggled on.

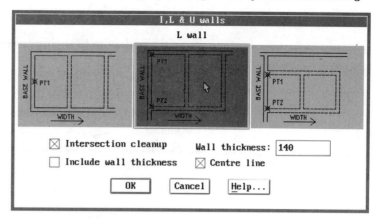

Fig. 3.17

5   Prompt: Indicate offset direction: pick any point to the right-hand, internal side of the base wall (Fig. 3.17).
6   Prompt: Space width: **3400** enter.
7   Prompt: Space width (or RETURN to exit): enter.
    Comment   Room widths can be chained for multiple insertion.

The internal partition wall should now appear on screen as shown in Fig. 3.18.

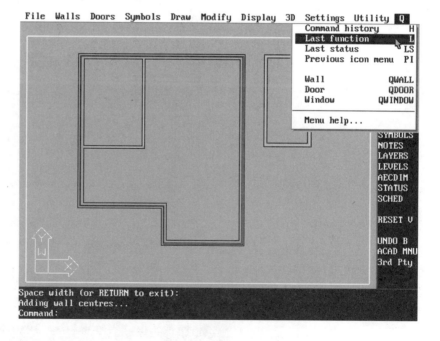

Fig. 3.18

Adding bedrooms

To add a bedroom in the upper right-hand corner of the building:

1   Pick [H4–5] on the tablet or Last function from the Q pull-down menu as shown in Fig. 3.18.
    Comment   Type L on the keyboard to call Last function if its easier.
2   Repeat the stages of forming the lounge (above) to insert the bedroom as shown in Fig. 3.19 (use dimensions of 4800 × 2600).

To form a second bedroom between the other two rooms

3   Pick [M4] on the tablet or choose I, L & U walls ... from the Walls utilities option on the Walls pull-down menu.
4   Pick I wall from the dialogue box as shown in figure Fig. 3.20.
5   Prompt: Pick point along base wall (pt2) <nea>: pick point as shown the dialogue box in Fig. 3.20.
6   Prompt: Space width: **3700** enter.
    Comment   You could also drag the cross hairs to this point if Snap were toggled on.
7   Prompt: Space width (or RETURN to exit): *enter.*
    Comment   Room widths can be chained for multiple insertion.

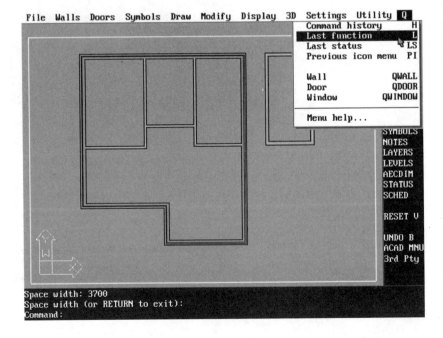

Fig. 3.19

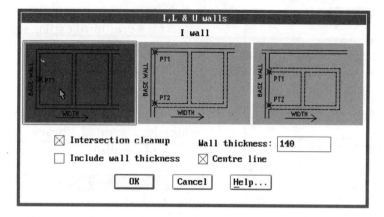

Fig. 3.20

The room should now appear on screen as shown in Fig. 3.19. Other rooms can now be added to form the kitchen and bathroom at this stage or can be left until later in the chapter when insertion of kitchen and bathroom fittings will include these wall elements.

Note   If you think that the larger bedroom is to narrow then stretch it for more space around the bed. Insert the furniture first from the domestic furniture options to see how much space is needed.

3.3 Bay window and special door insertions

Bay windows can be inserted into walls in 2D or 3D with one of four standard styles and using any dimensions you care to set up. Structural openings can be made at the same time or the wall break feature can be disabled.

Practical

During this session you will practise the following:

- Insertion of a full height square bay into the lounge wall, which could be used as a French window or have fixed glazing.
- Insertion of a circular bay into the front elevation to form a more open aspect to the dining room or as an area for indoor plants.

Inserting a square bay

1   Pick [D7] on the tablet or Bay windows ... from the Windows pull-down menu as shown in Fig. 3.21.
2   Select the Square bay style from the dialogue box as shown in Fig. 3.22.

Fig. 3.21

Fig. 3.22

Fig. 3.23

3   Select the other parameters as shown in Fig. 3.22 and refer to Fig. 3.23 for a graphical explanation of the chosen options.

Comment   Structural opening base height and bay sill height must be the same.

4   Pick the Name … text box and type BUNG10 (or choose another name) in the box as shown.

Comment   Once a bay has been defined it can be used as a standard symbol library item and inserted quickly into any type of dwelling.

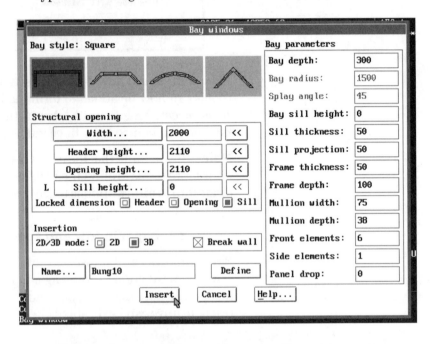

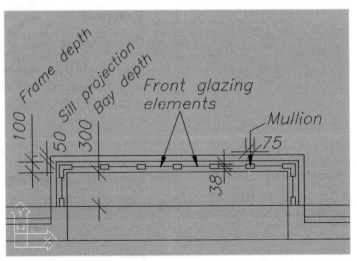

5 Click on Define to preview the bay window you have just created.

Comment   The shadow outline of the bay will be shown as a ghosted line on screen and can be cancelled if alterations are needed.

6 Click on Insert if satisfied with the design of the bay.

7 Answer the prompts to insert the bay by its midpoint at the wall midpoint.

Comment   Standard insertion prompts are used to place the bay.

Inserting a circular bay in the front elevation

1 Pick [D7] on the tablet or Bay windows ... from the Windows pull-down menu as described above.

2 Select the Circular bay style from the dialogue box as shown in Fig. 3.24.

3 Select the other parameters as shown in Fig. 3.24 and refer to Fig. 3.25 for explanation of the meanings of chosen options.

Comment   Structural opening base height and bay sill height must be the same.

4 Pick the Name ... text box and type BUNG11 (or choose another name) in the box as shown.

Comment   Now the bay has been defined it can be used as a standard symbol library item and inserted quickly into many different types of dwelling.

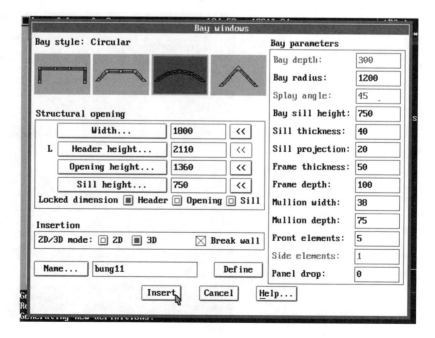

Fig. 3.24

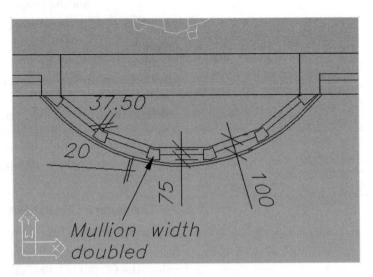

Fig. 3.25

5   Click on Define to preview the bay that you have created.
    Comment   The shadow outline of the bay will be shown as a
    white line on screen and can be cancelled if alterations are
    needed.
6   Click on Insert if satisfied with the design of the bay.
7   Answer the prompts to insert bay by its midpoint at the wall
    midpoint as shown in Fig. 3.25.

Both the inserted bays are shown in Fig. 3.26. Note that the dining
room has been formed by adding a partition wall.

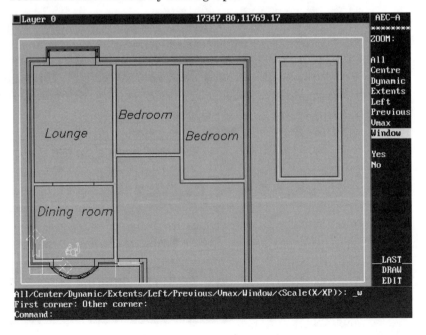

Fig. 3.26

Special doorways or windows can be inserted into any wall but you must design any frames that might be needed for insertion as user-defined symbols.

To form an open archway

To form an archway between the lounge and dining room as shown Fig. 3.27:

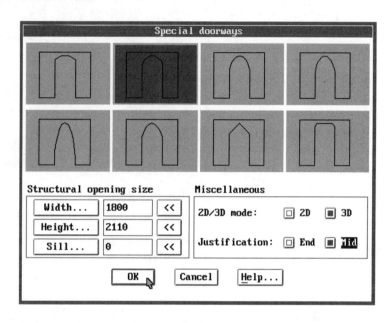

Fig. 3.27

1 Pick [B6] on the tablet or Special doorways … from the Doors pull-down menu as shown in Fig. 3.28.
2 Select the dialogue box options and type in parameters shown in Fig. 3.29.
3 Insert th doorway at the midpoint of the partition wall between lounge and dining room.

View in 3D on screen as shown in Fig. 3.27 using the perceive option.

To complete the front elevation, ready for elevation enhancement work in Chapter 4, add a front door and kitchen window to give a plan view of the 3D model as shown in Fig. 3.30.

Insert the front door with its swing fully open as previously described, then:

4 Pick [D5–6] from the tablet or Edit door style … from the Door styles option on the Doors pull-down menu, as shown in Fig. 3.31.

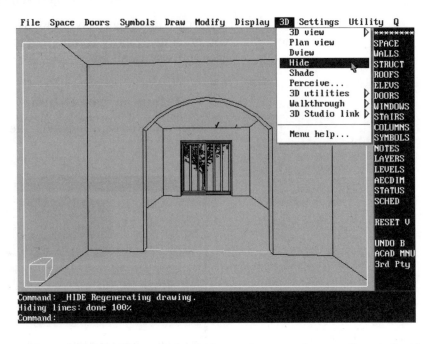

Fig. 3.28

5 Use the Edit door style dialogue box to change the style if required and close the door as shown in Fig. 3.32.

A hidden-line 3D view from a random Vpoint is shown in Fig. 3.33 with the front door closed.

Note   The wall convert command has been used to remove lines

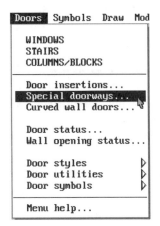

Fig. 3.29

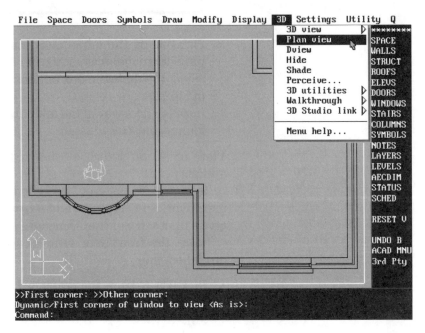

Fig. 3.30

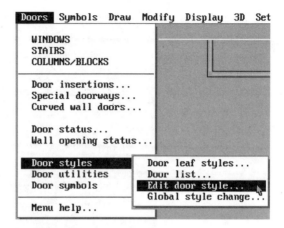

Fig. 3.31

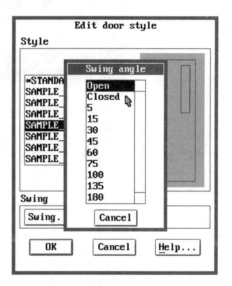

Fig. 3.32

above and below window openings and, as you can see, has worked on 80% of the 3D entities. Any lines left can be removed when 2D elevations are taken from the 3D model.

3.4 Roof construction options

These features of the software first appeared in AEC Version 2.0 as a series of limited options for the modelling of simple roofs. Version 3.0 contained a very extensive and completely rewritten series of options for the modelling of complex roofs of almost any shape and complexity. Many of the roof edit functions that caused difficulty for users of Version 2.0 were also automated in Version 3.0 allowing roof planes to blend at intersections. The manual edit routines of

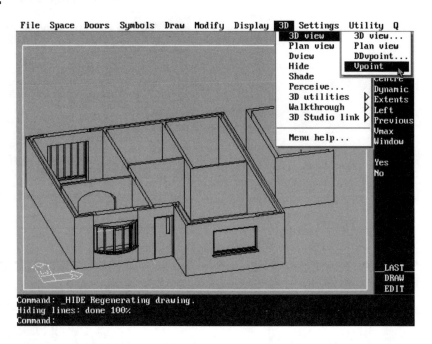

Fig. 3.33

Version 2.0 were retained for use in Version 3.0 but they are seldom required in practice. Gable walls were also introduced in Version 3.0 but they could not be pierced for window openings without using complex advanced 3D modelling techniques. AEC Version 4.0 has bettered the ease of use of the previous functions and has preview options to allow mistakes to be identified and corrected before placement of the roof takes place. Much more complex design work is possible than may be covered in a book of this length.

This section starts with the essential knowledge required in setting up viewports on screen as at least two views of the model are required for roof placement. The simple development of single-plane roofs and gables for the single garage is then covered. Lastly, hipped and gable roof options are used to form the main sections of the bungalow roof which are blended to form one composite element. As every good bungalow designer knows gable ends are very useful if you add a first floor which needs natural daylight and the lack of a standard gable window insertion feature in AEC is a major drawback.

If you are afraid of damaging the 3D model, try placing a pitched roof on the bungalow model using the AutoCAD Copy command to make a copy and practice on the copy first. In the example below, the bungalow or the single garage can be copied very quickly so that a variety of pitched roofs, not just the single-plane roof, can be tried and compared.

Practical

During this session you will practise the following:

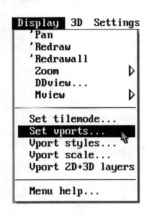

Fig. 3.34

- Setting up viewports to help insert gable and roof elements.
- Defining and placing gable ends to the garage and bungalow.
- Placement of a single-plane roof element on the garage.
- Placement of a multi-element roof on the bungalow.
- Adding eaves, soffit and other details to your own choice.

Setting up viewports

1   Pick Set tilemode ... from the Display pull-down menu as shown in Fig. 3.34.
2   Click the On option in the dialogue box as shown in Fig. 3.35. *Comment*   If you set the tilemode to 0 (Off) the Metaview port dialogue box appears together with the Paper space icon. This option is for plotting multiple views on hard copy and will be covered later.

Fig. 3.35

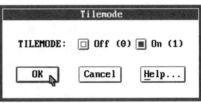

3   Pick [M6] on the tablet or Set vports ... from the Display pull-down menu as shown in Fig. 3.34.
4   Pick the two-viewport option as shown in Fig. 3.36.

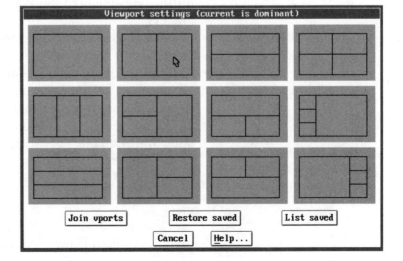

Fig. 3.36

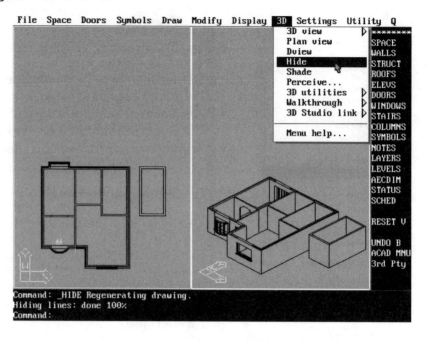

Fig. 3.37

Two plan views of the bungalow will now appear on screen so click on the right-hand viewport to make it active and set up an isometric view as shown in Fig. 3.37.

Note The text layer has been turned off to aid visualisation. Viewports are made active by pointing and clicking on them.

Gable walls

1 Pick [M3] from the tablet or Gable wall ... from the Wall utilities sub-menu of the Walls pull-down menu, as shown in Fig. 3.38.
   Comment Use a Zoomed isometric view to assist gable placement.
2 Pick the options in the dialogue box in Fig. 3.39 and click OK.
   Comment The dialogue box setup includes: mono-pitched structural single wall 200 mm thick; Alignment set to right and no Centre line; Eaves height of 2600 mm to leave no gap above other wall; End pitch of 35 degrees.
3 Prompt: From point: pick point 1 as shown in Fig. 3.40 using OSNAP END.
4 Prompt: To point: pick point 2 as shown in Fig. 3.40 using OSNAP END.

The gable end should now appear on screen as shown in Fig. 3.40.

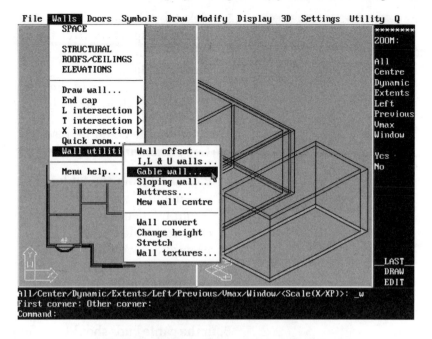

Fig. 3.38

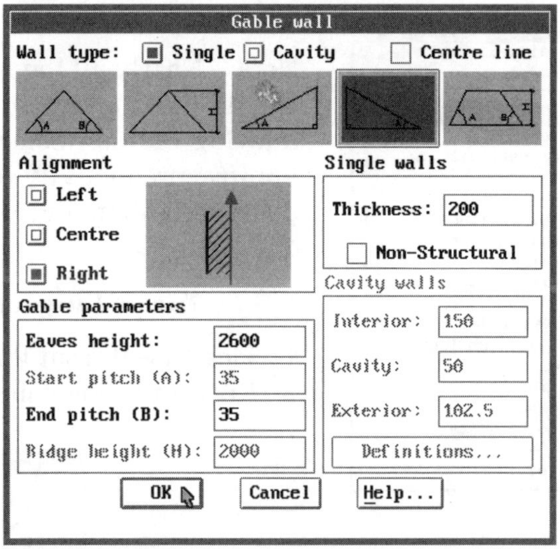

Fig. 3.39

Note The sequence of picking points must be handed the same way as the gable to ensure correct wall alignment and avoid an overhang.

You will probably have noticed by now how often you have been using the OSNAP END option and you may be wondering if the Snap could be set to END automatically; the answer is yes, and it is easy to set up.

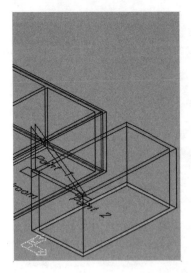

Fig. 3.40

Fig. 3.41

Fig. 3.42

5 Pick Object snap ... from the Settings pull-down menu as shown in Fig. 3.41 or from the tablet [T–U10].
6 Click on Endpoint in the select settings box to toggle it on as shown in Fig. 3.42.
   Comment   The size of the box used to pick Snap points can be set using the slide bar in this dialogue box.

The other gable could be inserted in the same manner but using the Copy command will probably be slightly quicker.

7 Pick [X15] from the tablet or Copy from AutoCAD commands sub-menu of the Modify pull-down menu as shown in Fig. 3.43.
8 Prompt: Select objects: pick Crossing from screen menu and cross the gable lines.
9 Prompt: Select objects: *enter.*
10 Prompt: <Base point of displacement>/multiple: pick one inner end of the gable.
11 Prompt: Second point of displacement: pick the outer end wall.

Both the gable ends should now appear on screen as shown in Fig. 3.44.

Note   Nearly all the AutoCAD Edit commands work in a similar way to the Copy command and may be used to modify existing drawing entities. Sometimes two edit commands are linked to improve productivity such as Copy and Rotate, covered in the site planning section, and Move and Rotate which is used all the time.

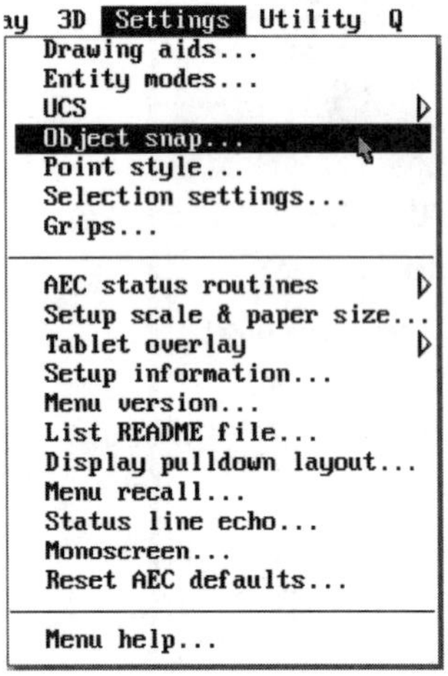

To roof the garage

1 Pick [A3–4] from the tablet or Roof constructions ... from the Roofs pull-down menu.

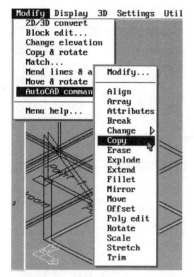

Fig. 3.43

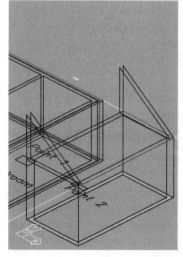

Fig. 3.44

2 Pick Single pitch roof from the Roof constructions dialogue box as shown in Fig. 3.45.

3 Prompt: Select the four corners of the roof:
Prompt: Box/First point: **B** *enter.*

Now activate the plan viewport as shown in Fig. 3.46.

4 Prompt: First corner of roof: pick corner of roof at point 1 as shown in the left-hand screen of Fig. 3.46.

5 Prompt: Other corner: pick corner of roof at point 2.
Comment   Use the Zoom window command before picking points then Zoom previous to see the model full screen once more.

6 Set up the eaves height and other parameters as shown in the dialogue box section of Fig. 3.47 as shown.
Comment   A clearance distance of 25 mm has been left above the wall tops to prevent the walls piercing the roof plane when viewing or plotting; choose your own dimensions for overhangs and pitch if you prefer. Click on the Dims and Rotate boxes to see other wonderful animated effects.

7 Click on 3D preview to check the roof's appearance and make any changes necessary then click OK.
Comment   For the first time Version 4 has allowed AEC users to insert roofs with the confidence that only a visual preview can give. This feature should spread through the interface to cover all construction elements.

Now bargeboards, fascias and soffits can be used to finish the

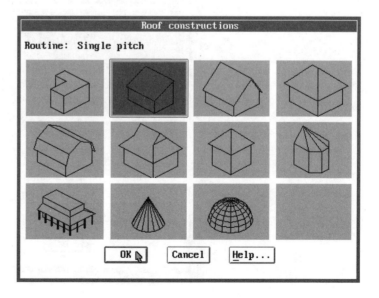

Fig. 3.45

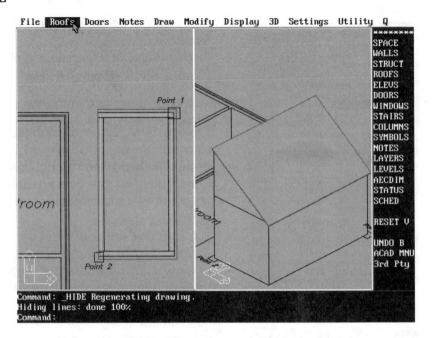

Fig. 3.46

single garage roof model to your own choice. The placement of
these elements has already been covered in the showroom modelling in Chapter 2 so refer back to that chapter if in doubt.

Note   A gap still exists in the 3D model on one side of the garage:
see if you can think of ways to fill that gap.

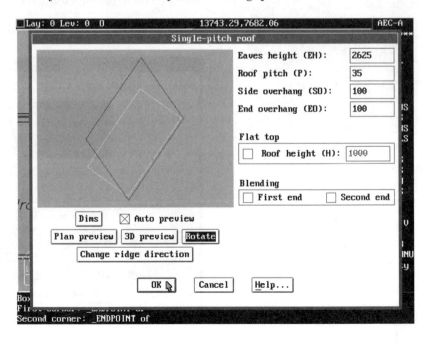

Fig. 3.47

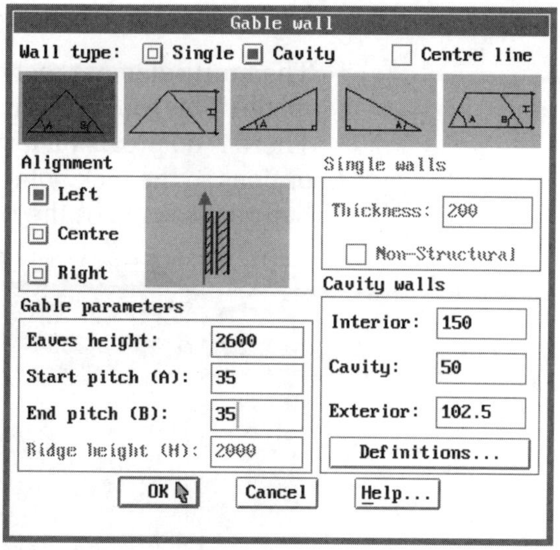

Fig. 3.48

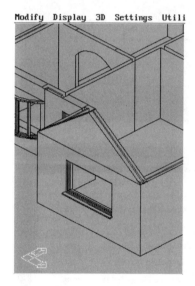

Fig. 3.49

Two more gables will now be added before the main bungalow roof is inserted. First set up the gable wall dialogue box as shown in Fig. 3.48. Set gable parameters exactly as specified. The previous cavity wall setting appears in the dialogue box automatically. Click on the Definitions box if you want to use a previously defined cavity wall. Insert two more gables which should appear as shown in Figs 3.49 and 3.50.

Note how the gable walls are all but invisible in the plan view but very prominent in the isometric view.

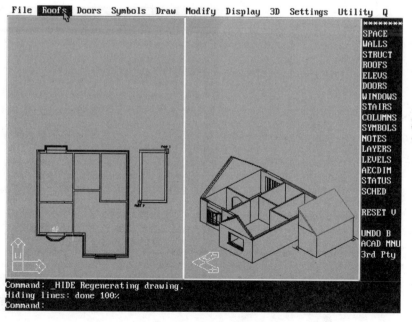

Fig. 3.50

Double-pitched gable and hipped roof to bungalow

The multi-element roof can now be placed on the main bungalow model. This roof will have a gable at one end and a hip at the other. There is no provision for an end overhang in the hipped roof setup options so the pick point for corner 1 of the roof must be provided before insertion of the roof element. The roof overhangs are very

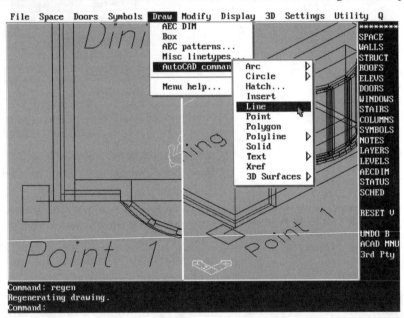

Fig. 3.51

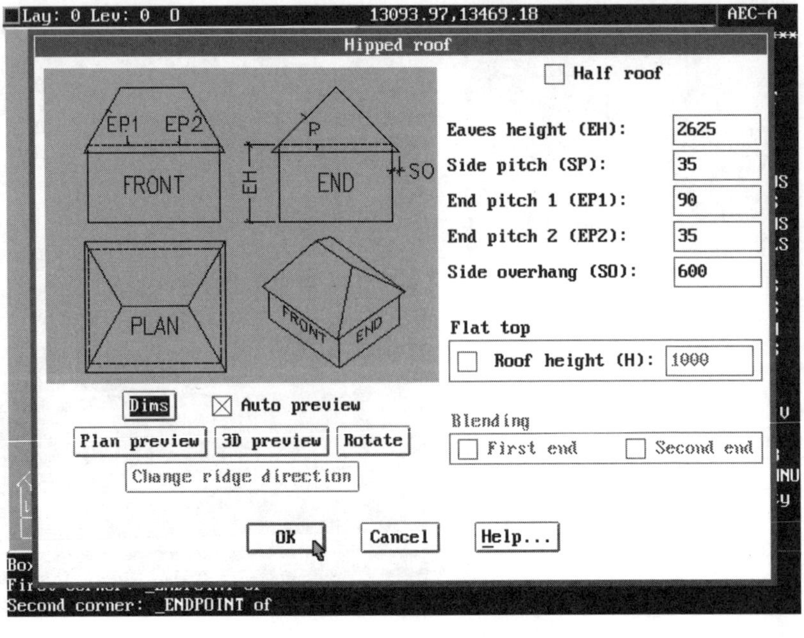

Fig. 3.52

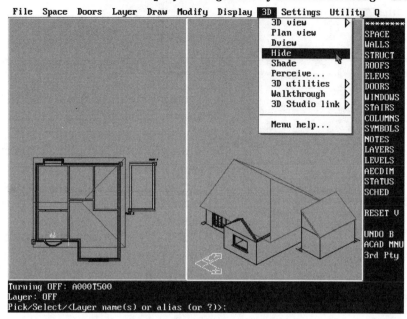

Fig. 3.53

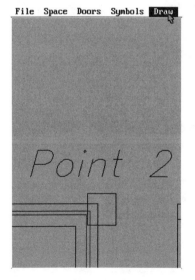

Fig. 3.54

important as they will provide covers to the bay windows and simplify the modelling required. Use the standard AutoCAD Line command to draw a single line 250 mm long and positioned as shown in Fig. 3.51. Toggle OSNAP END off before drawing the line.

First insert the main hipped roof:

1 Pick [A3–4] from the tablet or Roof constructions ... from the Roofs pull-down menu.
2 Pick hipped roof from the roof constructions dialogue box to bring the dialogue box shown in Fig. 3.52 on screen; choose the settings shown.

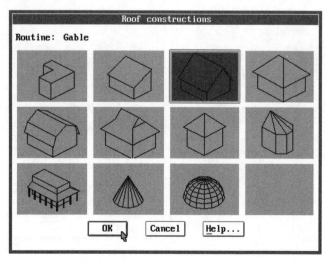

Fig. 3.55

File **Roofs** Doors Notes Draw Mo

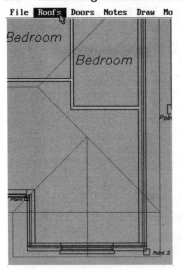

Fig. 3.56

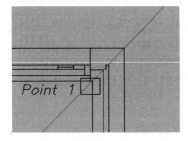

Fig. 3.57

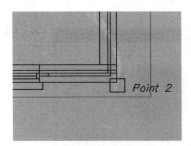

Fig. 3.58

Fig. 3.59

3   Prompt select the four corners of the roof. Prompt Box/First point: **B** enter.

Now activate the plan viewport as shown in Fig. 3.53.

4   Prompt: First corner of roof: pick corner of roof at point 1 as shown in Fig. 3.51.
    Comment   Pick end point of the single line as point 1.
5   Prompt: Other corner: pick top right corner of roof at point 2 as shown in Fig. 3.54.
    Comment   Use the Zoom window command before picking points then Zoom previous to see the model full screen once again.
6   Pick the Dims option in the dialogue box and set the height and other parameters as shown in Fig. 3.52.
    Comment   A clearance distance of 25 mm has been left above the wall tops to prevent the walls piercing the roof plane when viewing or plotting; choose your own dimensions for overhangs and pitch if you prefer.
7   Click on 3D preview to check the roof's appearance and make any changes necessary then click OK.

The roof should now appear in both viewports as shown in Fig. 3.53. Now add the smaller roof with gable end.

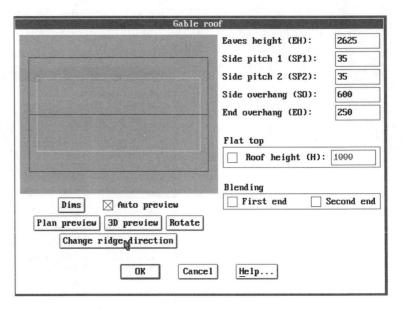

8   Pick [A3–4] from the tablet or Roof constructions ... from the
    Roofs pull-down menu.
9   Pick Gable roof from the Roof constructions dialogue box as
    shown in Fig. 3.55.
10 Prompt: Select the four corners of the roof.
    Prompt: Box/First point: **B** *enter.*

Now activate the plan viewport as shown in Fig. 3.56.

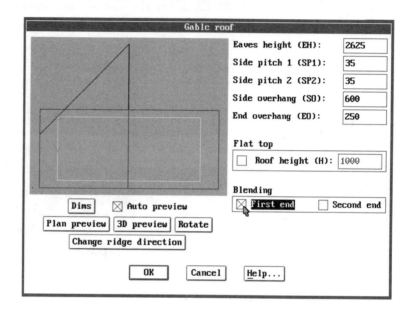

Fig. 3.60

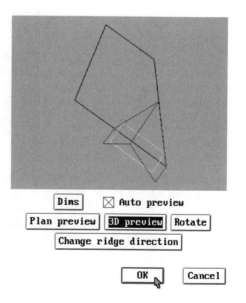

Fig. 3.61

11 Prompt: First corner of roof: pick point 1 at corner, as shown in Fig. 3.57.

12 Prompt: Other corner: pick point 2 at corner as shown Fig. 3.58.

13 Set up the eaves height and other parameters as shown in the Gable roof dialogue box section of Fig. 3.59.

14 Click on Change ridge direction as shown in Fig. 3.59.

15 Click on Blending for the First end of the roof as shown in Fig. 3.60.

16 Click on the Dims, Plan preview and 3D preview boxes as shown in Figs 3.61–63.

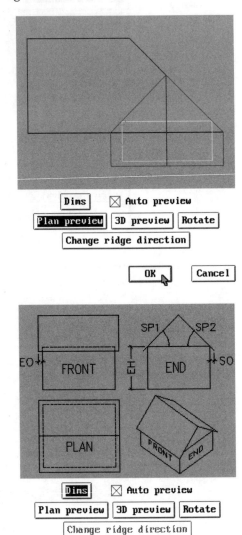

Fig. 3.62

Fig. 3.63

However, you will notice that one line remains visible between points 2 and 4 which should not be seen so use the Multi-sided roof command to place a Polymesh roof [B4] on this side of the building, or use the AutoCAD Hide edge command.

Now fascias and bargeboards can be used to finish the bungalow model roof as described for the garage pitched roof.

3.5 Symbol insertion used to design a fitted kitchen

To finish the 3D modelling on a lighter note design a fitted kitchen for the bungalow. You will need to complete the partition walls for the kitchen and bathroom first. Then follow the insertion sequence outlined below to get some knowledge of the methods used.

To insert 2D and 3D symbols:

1   Pick [F14–15] from the tablet or Kitchen fittings from the Symbols pull-down menu as shown in Fig. 3.64.
    Comment   Always use the library option on the pull-down to see a preview of the available symbols.
2   Pick unit F from the on-screen selection as shown in Fig. 3.65.
    Comment   There are often other options available on other menus such as wall units.
3   You could click on OK in the Unit width dialogue box to insert the standard unit but to see one of the most innovative features of AEC just click on the edit box instead as shown in Fig. 3.66.
    Comment   This is your first excursion into visual programming. I am sure that you will recognise the power of this feature to customise sizes of any inserted symbol, with a selection of your own sizes to choose from also.

Fig. 3.64

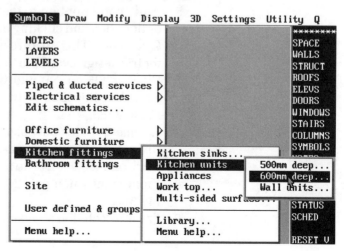

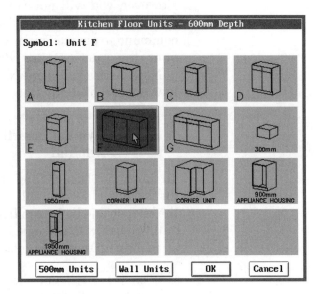

Fig. 3.65

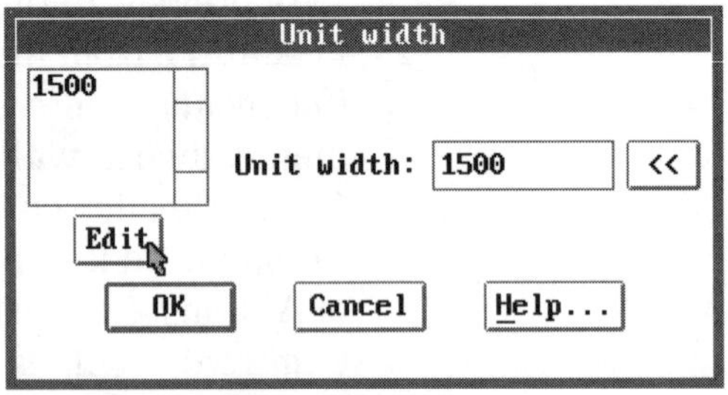

Fig. 3.66

4   A pop-up scroll box appears on screen.
5   Add another option to the list by typing 1800 in the blank New options box and clicking on Add new as shown in Fig. 3.67.
    Comment   This feature shows the power of making parametric design available to users.
6   Click on OK to exit and the Unit width dialogue box appears again but this time with two options to choose from as shown in Fig. 3.68.
    Comment   This new size option is now a permanent addition to the progam until you care to change it.
7   Click on 1800 not 1500 to insert the wider unit.
8   Prompt: Insertion point: insert at midpoint of wall as shown in Fig. 3.69.
9   Prompt: Rotation angle <0>: **180** enter.

Fig. 3.67

Fig. 3.68

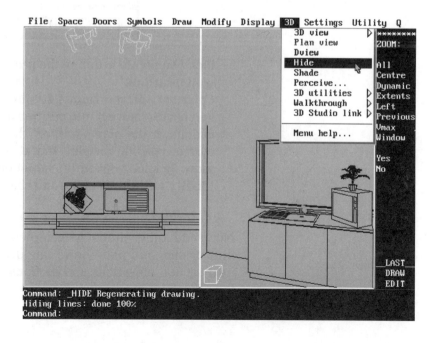

Fig. 3.69

Next select and insert the sink from the large selection included in the screen menu. When the Symbol insertion dialogue box appears on screen type in the correct height as shown in Fig. 3.70 and insert the sink.

The television and plant finish off your little group of inserted objects; they rest, of course, on the worktop.

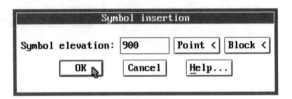

Fig. 3.70

To insert the worktop:

1   Pick Worktop ... from the Kitchen fittings pull-down menu as shown in Fig. 3.64.
2   Select the setting shown in Fig. 3.71 from the dialogue box and click OK.

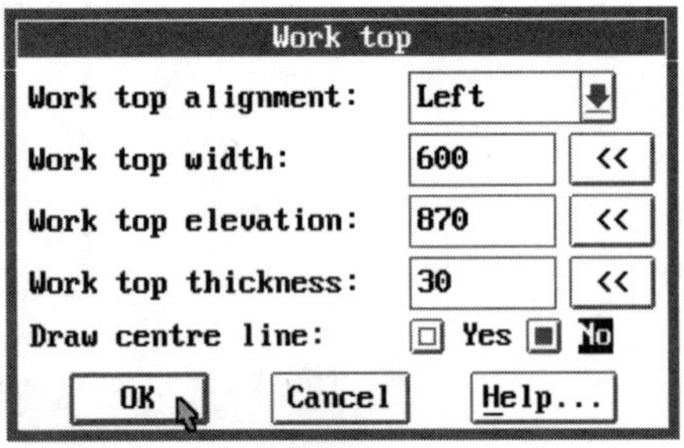

Fig. 3.71

3   Pick two points to define the position of the worktop just like a wall insertion.

Comment   If you cannot fix the worktop to points first time and have to move it into position do it in the 3D viewport. If this is done in the plan viewport the worktop will appear in the correct position on plan but at floor level on the elevational view. Pick [G12–13] on the tablet or Change elevation ... on the Modify pull-down menu and change the height.

Practice

Now add other units to complete the fitted kitchen. Three-dimensional and plan views can already be produced and in the next section elevations and sections will be added to the repertoire.

CHAPTER 4

Generating 2D elevations

Once a 3D model has been constructed the 2D elevations and sections can be created as separate drawings using a unique AEC feature which often astonishes CAD users who work with basic graphics engines like AutoCAD. Elevation generation shows views as 2D entities produced from a data file which is sorted to produce only those lines on screen that would be seen from a particular viewpoint. Any 3D model is literally interrogated from one designated side to see which lines would show from a particular angle. The only constraint is that the views must be orthographic (at 90 degree increments). The introduction of 'paper space' into AutoCAD allowed many views of a complete 3D model to be generated and plotted on the same drawing sheet. Any changes necessary could be made to the 3D model and would then automatically be reflected in the elevations when the drawing was plotted. Each version of AEC adds greater 3D functionality and the program's 2D features are gradually losing their importance. However, where automatic insertion routines are not available, working directly on detailed modifications to the various 3D faces of the model needs much practice before the process is fully understood and the interim method of generating elevations used by AEC is quick and easy. Always remember that elevation generation is a one-way process and that later modifications made to the elevations are not relayed back to the 3D model. This feature can be used on any of the models created in this book.

In this Chapter you will:

- Increase the size of your on screen working area to allow more views of the model to be shown.
- Use the wall convert option to remove unwanted lines from around openings.
- Generate elevations from the 3D model of the bungalow.
- Generate sections from the 3D model of the bungalow.
- Enhance elevations using RIBACAD and AEC elevation symbols.
- Use the AEC hatch patterns provided with the program.

4.1 Using setup to make more room

Before the production of elevations space must be made on the screen for placement of the elevations and sections. Starting new projects on a screen representation of A3 paper makes it very easy to expand the size of the drawing sheet on screen in several stages if necessary. To change the drawing sheet size on screen used to be easy, but with AEC Version 4 it became slightly more complex:

1   Pick [H22–23] from the tablet or Setup scale & paper size ... from the Settings pull-down menu as shown in Fig. 4.1.
2   Click on the Define style box as detailed in Chapter 1.
    Comment   Refer to the setup routines for earlier projects if you have any problems here.
3   Click on A1 sheet size and on OK to exit.
    Comment   The screen size can be changed to A0 later if required.

The screen is now redrawn to A1 size.

4.2 Wall convert

First try to make sure that the lines above and below window and door openings are removed from the 3D model; it is otherwise a long and tedious business erasing them from complex elevations

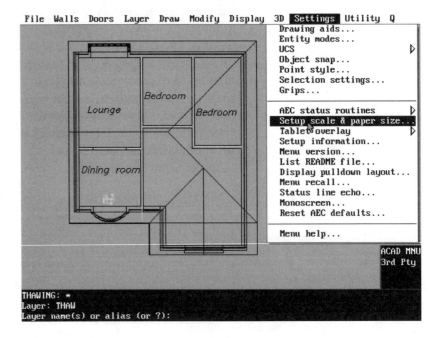

Fig. 4.1

later. Pick the Wall convert command described earlier from the tablet or pull-down menu.

You will now see the wall lines flash briefly on screen as this process proceeds to convert the walls by hiding the edges of the 3D faces where doors and windows are inserted.

Note   The Wall convert process works by creating a block from the wall faces to try and prevent the original lines above and below openings from showing. Unfortunately it also prevents you from working on individual pieces of wall so only use it when all work on the model is completed. Use the Explode command to separate individual wall lines for editing purposes if necessary. Having to use the Wall convert feature is one of the most frustrating faults in AEC; some other software packages based on AutoCAD manage to punch holes in walls without leaving lines to remove later, so look for improvement in later AEC releases or go for a solid modelling solution to this problem. It is sometimes quicker to modify doors, windows and other non-structural elements once the elevations had been taken off. However, this procedure allows the elevations to be developed without reference to, and independently from, the 3D model, which is obviously undesirable.

Copying the model

You may want to make a copy of the 3D model before taking off elevations to avoid any possibility of corrupting the original. Use the Copy command as described in previous sections to do this.

Note   Before using any Edit command on the 3D model make sure that all layers are 'thawed' and turned on otherwise the model will become fragmented with disastrous results.

4.3 Generating elevations

Producing elevations from the 3D model is a semi-automatic process and sets up a detailed framework for future enhancement by 2D drafting. AEC includes all 3D structural elements and any joinery in this process thus reducing the need for later 2D editing of elevations.

1   Pick [F3] on the tablet or pick Elevation label ... from the Elevs pull-down menu as shown in Fig. 4.2.
2   Select the options shown in the dialogue box that appears, as shown in Fig. 4.3, and click OK.

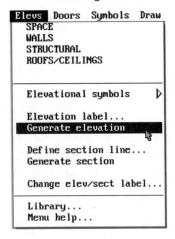

Fig. 4.2

3   Insert the tag in front of the elevation that you want to generate and select the viewing angle.

4   Pick [F1–2] on the tablet or Generate elevation … on the Elevs pull-down menu.

5   Prompt: Select elevation label: pick the elevation A tag that has appeared on the drawing.

6   Prompt: Select objects: **C** enter.

Comment   The generation time for elevations is directly related to the number of objects included in the crossing box so only select those that affect the elevation to reduce the data file size.

7   Prompt: First corner: Other corner: window all the 3D entities that will be seen from the elevation tag A.

8   Prompt: Pick basepoint: pick the bottom corner of the wall either inside or outside using an OSNAP as shown in Fig. 4.4.

Comment   If the eaves line is picked by mistake elevations will appear at the wrong base level so make sure the wall base line is selected.

Fig. 4.3

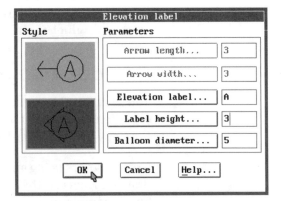

Fig. 4.4

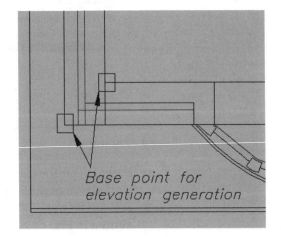

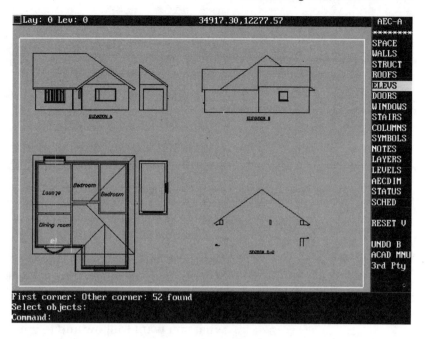

Fig. 4.5

9   Prompt: Pick translation point: toggle Ortho on and pick where
    to place elevation on the enlarged drawing sheet as shown in
    Fig. 4.5.
    Comment   The elevation will be directly above the plan if
    Ortho is toggled on and will not need repositioning later.

The elevation generator will now get to work and you will see
messages on screen in the following order: Writing data file;
Sorting lines; Processing lines. Speed of processing the information
depends upon your hardware, but is fast even on 386-class
machines.

10 Prompt: Reposition elevation <No> : *enter*.
11 The Generate another view dialogue box appears: click the No
    option to exit.

Practice task

Now take off another elevation from the garage side. Pick the
bottom corner of the vertically aligned right-hand side elevation
as the base point. As the elevation is generated it is automatically
turned round to the horizontal and the bottom corner becomes the
left-hand corner as shown in Fig. 4.5.

Generating sections

Sections are generated in the same way as elevations and show similar detail which must be fleshed out later using 2D drafting techniques. RIBACAD contains a large number of pre-drawn section details which are easy to insert and edit.

To take off a section

1 Pick [G3] on the tablet or Define section line ... from the Elevs pull-down menu then define the Section line as shown in Fig. 4.6.
2 Pick [G1–2] from the tablet or Generate section from the Elevs pull-down menu.
3 Position the section as described for the elevations (Fig. 4.1).

Note   Whether to use 3D modelling or 2D drafting techniques in construction design is the subject of constant debate amongst clients, designers, constructors and others in the industry. Like all complex questions there is no simple answer, it all depends on what you want to show and to whom. In the recent past little could be done with 3D hardcopy output by PC users but when cheap add-in rendering engines started to appear using the AutoCAD ADS system the situation was rapidly transformed and 3D became mainstream. A firmly established trend now exists to use 3D modelling and visualisation wherever possible as a general purpose design tool.

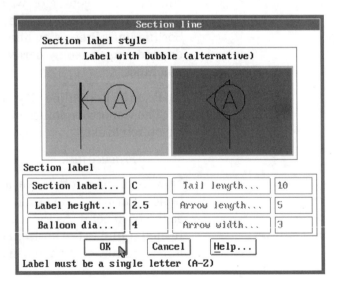

Fig. 4.6

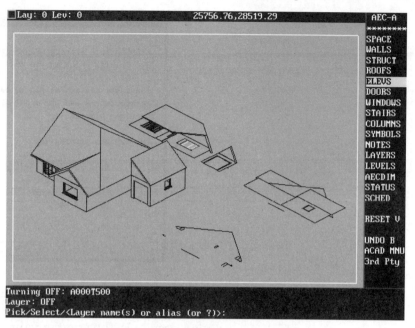

Fig. 4.7

Editing elevations and sections

As you can see in Fig. 4.7, 3D and 2D views now exist on the same screen. The 2D views can be edited using any standard AutoCAD Edit command to produce the required views and then enhanced by some insertion features of AEC.

A 2D floor plan of the 3D model can also be instantly created a by picking [H12–13] on the tablet or 2D/3D convert from the Modify pull-down menu and windowing all objects to be converted.

Note Plain wall sections are not converted and remain in 3D.

4.4 Enhancement of elevations and sections

Once the outline sections have been generated they can be enhanced and then completed by both RIBACAD and AEC routines to produce the finished detailed views.

RIBACAD electronic library

This is a unique CAD service, described in Fig. 4.8, for building and civil designers offered by RIBA Services Ltd from its London base.

It has grown dramatically from the initial few drawings provided by one or two manufacturers to a major resource in just a few years. There are now over 7500 pre-drawn 2D details available

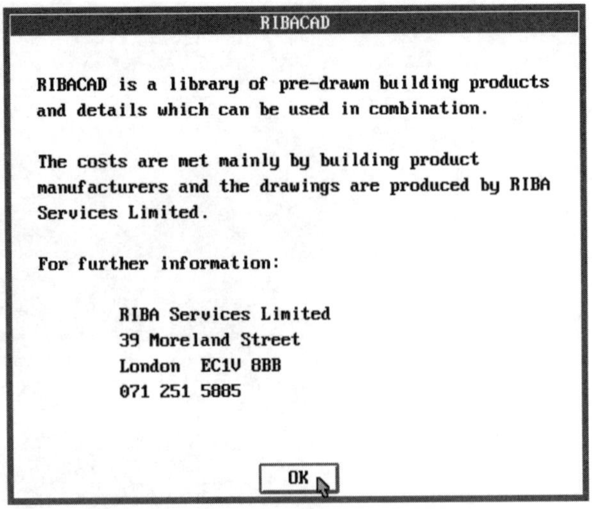

```
┌──────────────────────────────────────────────┐
│                    RIBACAD                      │
├──────────────────────────────────────────────┤
│                                                │
│   RIBACAD is a library of pre-drawn building products
│   and details which can be used in combination.│
│                                                │
│   The costs are met mainly by building product │
│   manufacturers and the drawings are produced by RIBA
│   Services Limited.                            │
│                                                │
│   For further information:                     │
│                                                │
│       RIBA Services Limited                    │
│       39 Moreland Street                       │
│       London  EC1V 8BB                         │
│       071 251 5885                             │
│                                                │
│                                                │
│                        ┌────────┐              │
│                        │   OK   │              │
│                        └────────┘              │
└──────────────────────────────────────────────┘
```

Fig. 4.8

from some 100 component manufacturers located world wide. RIBACAD offers the opportunity of incorporating these into AEC design drawings, thus saving valuable time at the detailing stage and also reducing errors which may occur if the designer misreads detail from the manufacturers drawings. Specification clauses are sometimes included with the details to reduce mistakes and the busy designers workload still further. There is nothing quite like it in the engineering or other CAD fields where nearly all detailed drawings are heavily guarded to protect propriety ideas (which are patented or copyrighted) from being passed on to potential competitors.

An exhaustive list of components is included in the library from bricks to bathrooms and from wardrobes to wall ties. Regular upgrade issues of RIBACAD are released to add material from an ever-increasing base of manufacturers to the growing library. New releases have added products such as mineral wool insulation, steel cladding, and roofing tiles. Several ranges of windows and doors made from timber, steel and PVC are included.

RIBA Services have provided a graphical interface for parts of the system which includes pull-down menus for rapid command input of some manufacturer's details. An icon-driven method of selecting drawings for insertion could be provided in the future. As a start to this process the Marley Tile Company have now issued their own addition to RIBACAD which allows complete sections of roofs to be inserted using a simple menu system which can be added to the AEC command system.

A full list of third-party software programs related to AEC is accessed from the third-party info option. As the numbers of floppy discs required for RIBACAD and other third-party draw-ings has grown storage has become a major problem. CD-ROMs are the cheapest method of storage now that the necessary hardware is widely available and the whole library is available in this format. High capacity floptical disks are another viable alternative with up to 25 Mb on one 3.5" floppy disk. Other methods are being introduced as new technologies come on stream. With the increas-ing use of 3D modelling the library should now have been extended to incorporate 3D symbols for items such as furniture and other appliances. Extensive documentation in several ring binders are provided to support the electronic information. The whole package represents very good value with only a small annual subscription required to secure the upgrades available each year.

The RIBACAD design process

Using RIBACAD increases your options when it comes to organisa-tion of the drawing process and there are many ways to get the job done faster using RIBACAD pre-drawn details.

One such might be as follows:

1 Draw a freehand sketch of the enhanced section required.
2 Find the items needed in the RIBACAD catalogue and note the RIBACAD file numbers on the freehand sketch.
3 Select the individual CD-ROM or floppy disks which hold the items needed.
4 Insert all the items required on screen and place them beside the generated outline section.
5 Use the AutoCAD Explode command then edit the RIBACAD details so that they fit inside the outline sections or elevations generated.
6 Add AEC hatch patterns or textures and text to complete the sections.

    Comment This may seem like a complex process but your assembly of components can then be blocked using the Auto-CAD Block command and written to disk for storage and rapid insertion into future drawings. See Chapter 7 (and any general AutoCAD text) for full details of blocks and their use. A library of blocks showing standard sectional details could be quickly built up based on standard specifications for rapid insertion.

As an example use a foundation section detail provide by Therma-lite Ltd:

7 Insert the floppy disk or CD-ROM into the necessary drive of the computer.

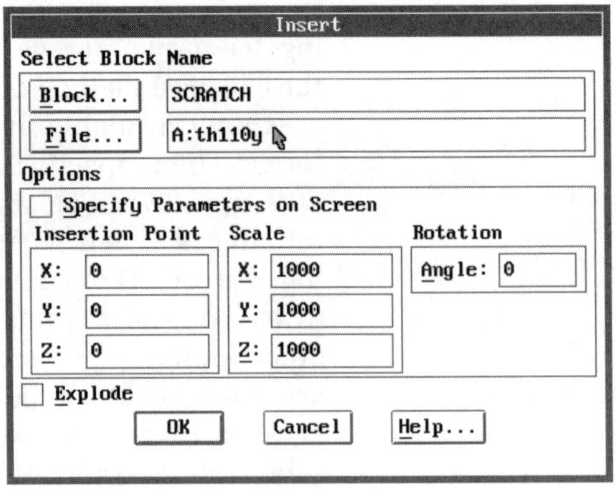

```
┌─────────────────────────────────────────────────┐
│                    Insert                        │
├─────────────────────────────────────────────────┤
│  Select Block Name                               │
│   ┌──────────┐  ┌──────────────────────────────┐ │
│   │ Block... │  │ SCRATCH                      │ │
│   └──────────┘  └──────────────────────────────┘ │
│   ┌──────────┐  ┌──────────────────────────────┐ │
│   │ File...  │  │ A:th110y ▷                   │ │
│   └──────────┘  └──────────────────────────────┘ │
│  Options                                         │
│   ☐ Specify Parameters on Screen                 │
│   Insertion Point   Scale           Rotation     │
│   X: ┌────┐   X: ┌──────┐      Angle: ┌────┐      │
│      │ 0  │      │ 1000 │             │ 0  │      │
│      └────┘      └──────┘             └────┘      │
│   Y: ┌────┐   Y: ┌──────┐                         │
│      │ 0  │      │ 1000 │                         │
│      └────┘      └──────┘                         │
│   Z: ┌────┐   Z: ┌──────┐                         │
│      │ 0  │      │ 1000 │                         │
│      └────┘      └──────┘                         │
│   ☐ Explode                                      │
│        ┌────┐   ┌────────┐   ┌────────┐           │
│        │ OK │   │ Cancel │   │ Help...│           │
│        └────┘   └────────┘   └────────┘           │
└─────────────────────────────────────────────────┘
```

Fig. 4.9

8 Pick [R8] from the tablet or Insert ... from the Draw pull-down menu.

9 Pick File and type **A:TH110Y** in the dialogue box space next to the file option as shown in Fig. 4.9.

   Comment   This is a Thermalite code number for a foundation detail. To speed multiple insertions just press the space bar and the dialogue box reappears.

10 Set up the other options as shown in Fig. 4.9 to insert the unexploded detail at the bottom left corner of the screen with a scale factor of 1000 and a rotation angle of 0.

   Comment   Click on the Explode option to explode the drawing block into individual lines that can be edited in detail. Only select this option if you can position it exactly on screen during insertion and it will not then be moved. Use scale factor of 1000 if using the RIBACAD symbols drawn in millimetres or accept the default value <1> if in metres.

11 Use the AEC Edit commands such as Move/Rotate and Copy/ Rotate, supplemented by other commands, to position the symbols on screen in their final positions.

4.5 Inserting features into elevations

Design philosophy

Enhancing generated elevations and section is a 2D process. At present if you change the model then that change is reflected in the elevations but not vice versa. However, this is only a transitional stage towards full interaction between the 3D model and 2D projections from the model. Some confusion can arise when generating other documents from the drawings. Window and door

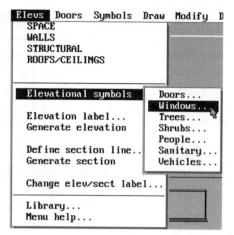

Fig. 4.10

schedules for example are generated from the insertions into the 3D model, not from those inserted into 2D elevations.

A wide range of windows and doors can be inserted into elevations, including all the commonly used domestic styles, with the opposite corners of the existing opening being used to fix the size of the component being inserted.

Casement window insertion

1   Pick [F4] from the tablet or Windows ... from the Elevational symbols sub-menu of the Elevs pull-down menu, as shown in Fig. 4.10.
2   Select the Casement/pivot option from the screen menu as shown in Fig. 4.11.
3   Prompt: Enter lower left corner of opening <END>: pick lower

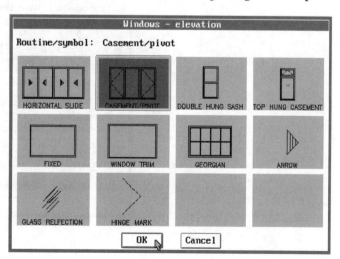

Fig. 4.11

Elev window – casement/pivot		
Glass panes (2/3/4):	3	
Side stile width:	38	<<
Top & bottom rail width:	38	<<
Reveal width:	20	<<
OK	Cancel	Help...

Fig. 4.12

left-hand corner of kitchen wall opening.

4   Prompt: Enter upper right corner of opening <END>: pick upper right-hand corner of kitchen wall opening.

5   Enter the windows parameters in the dialogue box shown in Fig. 4.12.

The finished window will now appear on screen.

Many other features can now be added such as hinge marks and reflection marks from the Windows pull-down menu. When inserting RIBACAD windows remember not to enter a scale factor but to fix window's position by picking first and second points with OSNAPs.

Door insertion is very similar to the routine for windows. Indeed, some of the doors selected seem to think that they are windows and give you the same prompts for insertion as windows. Many detailed features can be added to doors from the library accessed from the Elevs pull-down menu including letter boxes, furniture and street numbers. See Fig. 4.13 for some possible options.

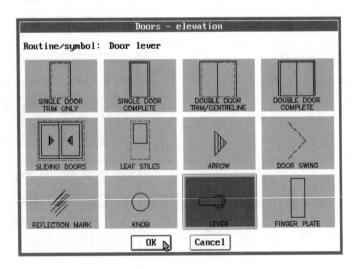

Fig. 4.13

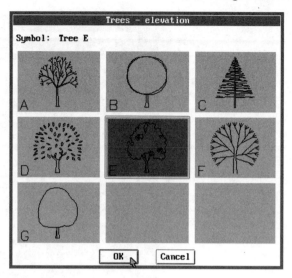

Fig. 4.14

Tree, shrubs and people

Various types of trees, shrubs and people can be selected from another screen menu to further enhance elevations. These are 2D drawings inserted as blocks and you will probably want to make your own and add to the ones provided in order to build up a personalised library of symbols.

See Figs 4.14 and 4.15 for some possible options.

Practice task

Add windows, doors, people and trees of different types and styles to complete one elevation. Try to choose the various symbols

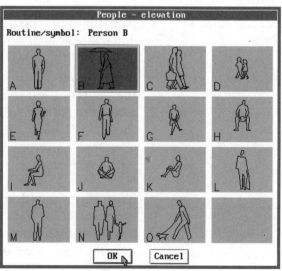

Fig. 4.15

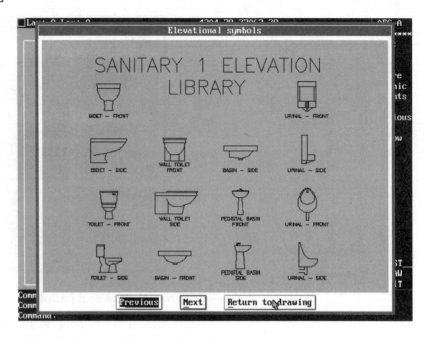

Fig. 4.16

to produce a harmonious balance.

Many other symbols available from the library are shown in Fig. 4.16.

Hatching elevations

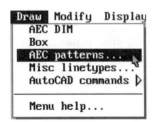

Fig. 4.17

The only enhancement to the standard but limited AutoCAD hatching routines are the addition of AEC elevation hatch patterns and texture patterns for walls and sections.

1   Pick [R9] on the tablet or AEC patterns ... from the Draw pull-down menu, as shown in Fig. 4.17.
    Comment   Textures for plans and sections can be accessed from [N4] on the tablet or from the Wall utilities sub-menu of the Walls pull-down menu.
2   Select the pattern AECREALBRICK1 from the screen menu as shown in Fig. 4.18.
    Comment   The options include patterns suitable for walls, roofs and floors.
3   Follow the standard AutoCAD hatch routines to insert the hatch patterns within defined boundaries.
    Comment   Use the AutoCAD hatch preview feature before inserting any hatch pattern to check that the scale is suitable for the area being hatched. AECREALBRICK1 will be inserted (at a scale factor of 1) into the defined boundaries. Other patterns

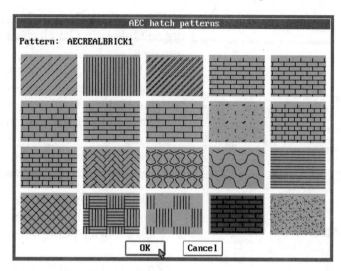

Fig. 4.18

can be chosen at different scales to suit the elevation or section being produced. AECINS for example can be used for roofs but a suitable scale factor will be somewhere between 200 and 300.

Note If you should select the wrong scale for hatch patterns AutoCAD will not be intelligent enough to relate the pattern to the shape being hatched. You could well end up with a solid fill pattern the takes many minutes to draw on screen. Never guess the scale and always use the hatch preview option first before final insertion.

Boundaries for the hatch can be defined by using the AutoCAD boundary hatch features applied to existing wall or window edges. However normal architectural practice and the need for small drawing file size is better served by defining a small area of wall or roof. Boundaries can be defined using standard AutoCAD Polyline and Spline curve techniques. Drawing file sizes can finish up being very large if you hatch large areas of wall or roof on elevational views.

4.6 Adding textures to plan views

All the major British and European standard hatch patterns are accessible on the Textures slide menu by picking [N4] on the tablet or Wall textures ... from the Wall utilities sub-menu.

User-created patterns can be inserted as blocks to augment those provided as standard, but production of such patterns is outside the scope of this work.

Hatch patterns scales are automatically set so that they fit the wall into which they are being inserted at the right spacing, unlike the AEC elevation hatch routines and the standard AutoCAD hatch routines.

Use the Zoom window and Zoom previous commands to avoid picking unwanted end points on long lengths of wall, and remember that picking any point on a line with an endpoint Snap will locate the endpoint nearest to the point picked.

Picking the nearest point on the wall line that is in the same half as the endpoint required but clear of other end points will solve this problem and avoid excessive zooming on long lengths of wall.

1   Pick [N4] on the tablet or Wall textures ... from Wall utilities on the Walls pull-down menu, as shown in Fig. 4.19.

2   Pick the brick pattern from the menu on screen as shown in Fig. 4.20.

3   Zoom window to enlarge a section of wall.

4   Prompt: Enter 1st corner of wall <End>: pick point 1 on horizontal wall.

5   Prompt: Enter 2nd corner of wall <End>: pick point 2 on horizontal wall.

6   Prompt: Enter 3rd corner of wall <End>: pick point 3 on horizontal wall.

7   Prompt: Enter 4th corner of wall <End>: pick point 4 on horizontal wall all as shown in Fig. 4.21.

8   Prompt: Hatch orientation Reverse/<Normal>: *enter.*

The hatched area should now appear on screen as shown in Fig. 4.22.

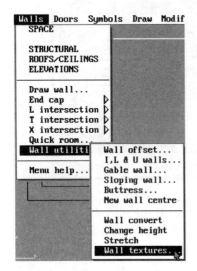

Fig. 4.19

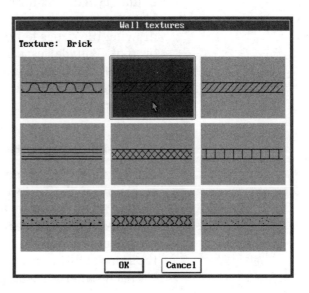

Fig. 4.20

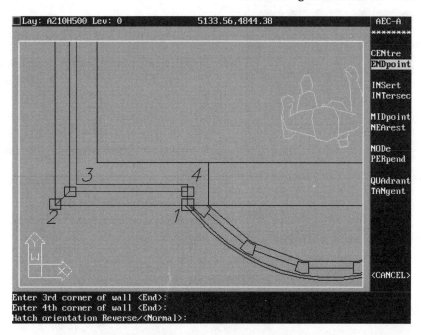

Fig. 4.21

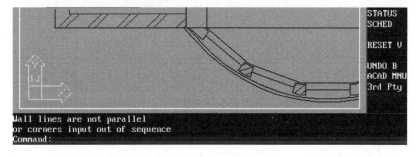

Fig. 4.22

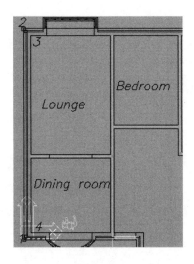

Fig. 4.23

Note Some sections of wall are too big to hatch in one length so use Zoom window and Zoom previous to locate important areas. As the OSNAP is set to END you can pick any point on the wall half nearest to the desired line end. When hatching, the wall length is defined by the first two picks, direction of picks may be clockwise or anticlockwise as desired.

Next hatch the adjoining vertical wall using the pick points shown in Fig. 4.23. Remember to reverse the hatch orientation for adjoining walls so that they match or use the match command if they do not.

The Hatching should now appear as shown in Fig. 4.23.

Hatching small areas

In productivity terms it isn't worth bothering to hatch small areas such as the cavity closers on the inner leaf of a cavity wall or small

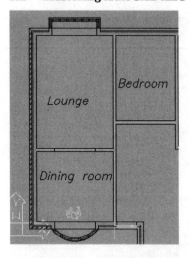

Fig. 4.24

stub walls. However, if you think hatching worthwhile in certain instances, the following tips may help:

- If you are determined to hatch every available section of wall it might be better to use the Belgian cavity detail and then add the closers later or just use the single wall.
- Often you will find that the hatch pattern does not match the rest of the wall in scale or direction, and to proceed further involves manipulation of the hatch pattern on a line by line basis.
- Refer to Section 4.4 on enhancement of 2D elevations for details of the commands needed.
- Only small areas need be hatched to show the type of material used.

CHAPTER 5

Adding notation to drawings

One of the major criticisms often levelled by some people at a CAD drawing is that it lacks the style of an individual hand. In engineering this is usually of no importance as the end user rarely sees the technical documents that describe a product. However the architect's or building designer's detail drawings of a project will often be submitted to the client for approval and the look and feel of a drawing will often sway the non-technical client's opinion about a proposal, regardless of its technical merits. A wide range of text fonts and styles are available in AEC and some are illustrated in Figs 5.1 and 5.2. Now that a Qtext (quicktext) option exists in the programme all those complex fonts that looked smart, but took hours to regenerate and plot, can be used. Use the options available in a skilful manner and no hand drawing will come close for artistic impression. Try producing A1 posters for a presentation in Gothic English with text of any height and, say, 15-degree inclined letters for the individual touch. You will be very surprised at the quality of production available, particularly with creative use of the Edit facilities to enhance borders with boxes and polylines.

Due partly to these requirements the text handling features found in AEC go far beyond the capabilities of AutoCAD, or indeed most other CAD packages, and should satisfy the most demanding

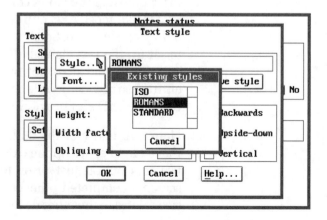

Fig. 5.1

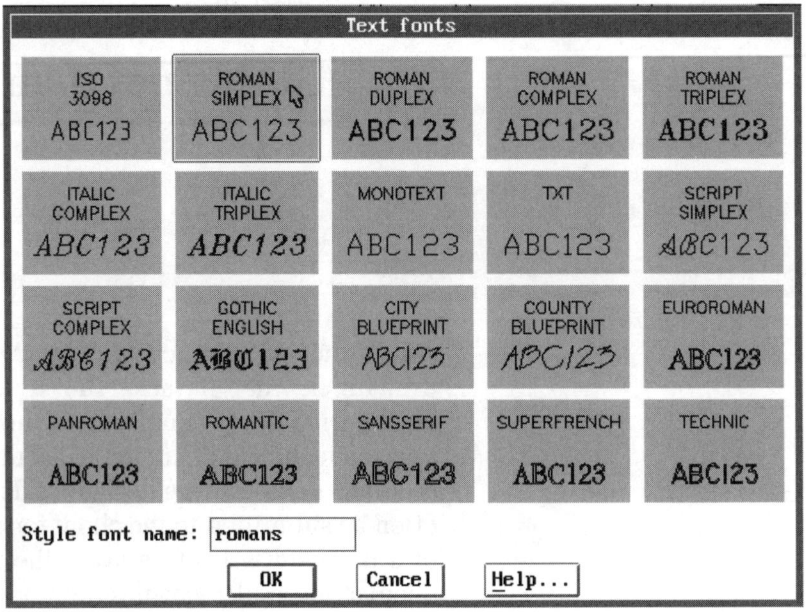

Fig. 5.2

of construction CAD users. The text edit and file linking routines represent a major advance for CAD packages and offer a foretaste of the integrated document-based software of the future. Always use AEC Notes rather than the crude AutoCAD text commands which have now been relegated to due obscurity in the command structure.

5.1 Notating drawings

As notes are used on all construction project designs this Chapter covers routines that could be used on any design image or document. All text used in the drawings produced for this volume were inserted using the Notes command with the settings shown in Fig. 5.1. Unlike most AEC entities text can be changed in many ways before or after insertion but it will usually only take a few seconds to set up the text status before insertion. Do not forget that text is strictly 2D and does not obey the same rules as other drawing entities, so always double check its final form before plotting using the AutoCAD Plot preview command.

Note   In many of the dialogue boxes illustrated you will notice that a Paper space option exists to insert notes in Paper space. Only use this option just prior to plotting when everything else on screen is completed; otherwise insert text in Model space so that its size relates to the other drawing entities.

Practical

During this session you will practise the following:

- Setting up text fonts, styles, sizes and other parameters.
- Entering text of various types.
- Editing text of various types.
- Applying tags and labels to the drawing.

Setting up text styles

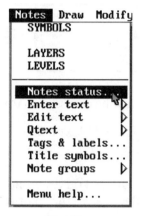

Fig. 5.3

1 Pick [E21] from the tablet or Notes status ... from the Notes pull-down menu as shown in Fig. 5.3.
2 Click on Set style ... in the Notes status dialogue box that appears on screen as shown in Fig. 5.4.
3 Enter the options shown in the Text style dialogue box as illustrated in Fig. 5.5 and click OK to exit.
   Comment   Save the style for later use. Do not change the text height from 0, rather use AEC Notes to set size as in step 4 below.

Fig. 5.4

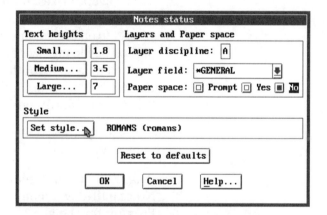

Fig. 5.5

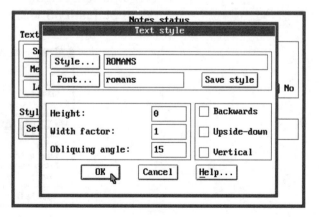

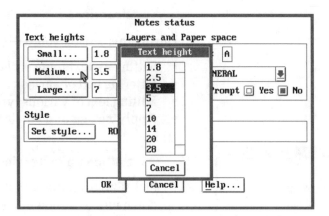

Fig. 5.6

The main Notes status dialogue box can now be accessed again.

4   Click on the Small …, Medium … or Large … text height boxes
and set up the new text heights you require for note entry as
shown in Fig. 5.6.
Comment   This allows for a standard format on drawings that
need several sizes of text. In Model space the plotted or
displayed sizes will vary relative to the graphical entities. If
you set to on the Paper space option the sizes chosen will be the
plotted hardcopy sizes and unrelated to the graphical entities.

5   Set the Paper space option to No.
Comment   Notes in Paper space must be entered after everything else on the drawing is finished.

Having set up the style you can now insert the medium text
anywhere on screen.

6   Pick [F19] on the tablet or Enter text, Medium text from the
Notes pull-down menu as shown in Fig. 5.7.

7   Prompt: Text height: 3.5 mm.
Prompt: Justify/Style/Start point: Text: pick a point on screen

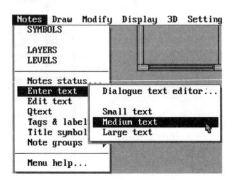

Fig. 5.7

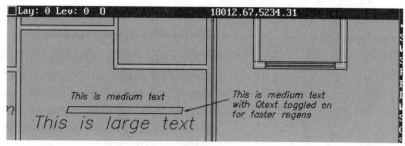

Fig. 5.8

where you want the text to start.

Comment A small white box should appear on screen where the text will start.

8 Type in 'This is medium text' as shown in Fig. 5.8.

Comment Just press *enter* when you want to start a new line below the first, or pick a point elsewhere to start some new text.

9 Press *enter* then *enter* again to leave command.

Comment Do not forget the second *enter* which exits the text command.

Practice task

Add to your drawing by picking the Large text option and typing 'This is large text' as also shown in Fig. 5.8.

Note Text for specifications or longer descriptions can also be added through the dialogue text editor as shown in Fig. 5.9. Use it

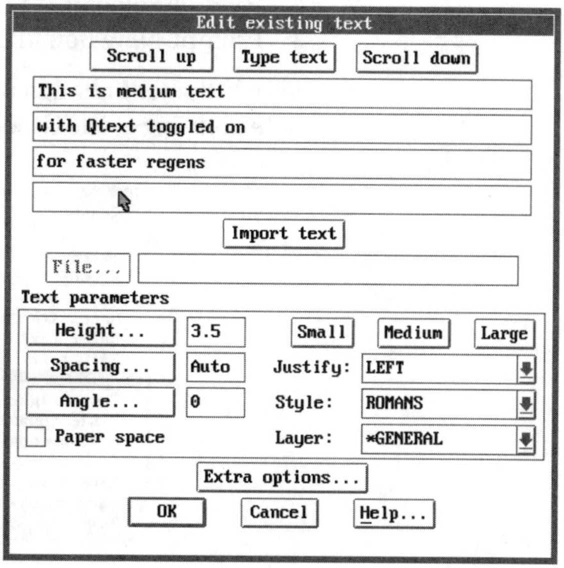

Fig. 5.9

like a simple word processor; note however that copy, cut or paste commands are not supported.

Text can also be entered 'Greeked' (represented by just a box on screen) for rapid screen regeneration by using the Qtext toggle as shown in Fig. 5.8. You will find this option invaluable for drawings that include specifications or large quantities of revision notes, particularly as it can be applied selectively or globally to any text before or after entry.

To insert Greeked text

1   Pick [H20] from the tablet or Qtext from the Notes pull-down menu as shown in Fig. 5.10.
2   Pick Set Qtext ON from the cascading menu.
3   Enter any text by the methods already described.

Note   Text previously entered as described above can be shown in full or Greeked by picking the Regen selected text command (Qtext OFF or Qtext ON options) and then selecting the text to be converted.

Editing text

Next you can edit some text, first by just changing the height then more comprehensively.

1   Pick [F21] from the tablet or Change height only from the Edit text option on the Notes pull-down menu.
2   Prompt: Select text to edit ?: pick the medium text.
3   Prompt: New height: **7** *enter.*

Watch the text change on screen to the new height. Now edit the note pointing at the Greeked text seen in Fig. 5.8.

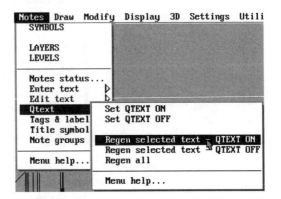

Fig. 5.10

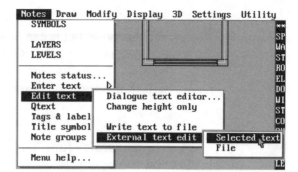

Fig. 5.11

4   Pick [E20] from the tablet or Edit text, Dialogue text editor from the Notes pull-down menu as shown in Fig. 5.11.

5   Make any alterations you wish to the text shown in Fig. 5.9 and click on OK to exit the box.

Comment   Note the wide range of options available to control the altered text. Just about everything except the font can be changed direct within this text editor.

6   The altered text will now appear on screen.

While you may still be reeling in amazement from such capabilities, there is yet another little piece of functionality to reveal.

7   Pick Edit text, External text edit from the Notes pull-down menu as shown in Fig. 5.11 and a fully fledged MS-DOS text editor will appear, as shown in Fig. 5.12.

8   Copy, cut and paste within the AECTEMP.TXT file created.

9   Save your text to a file and try opening it again within the text editor using the File menu.

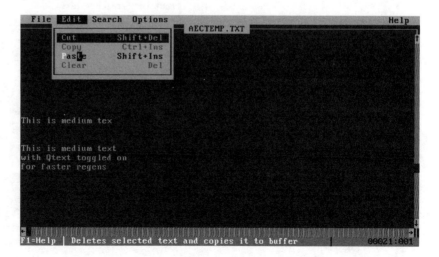

Fig. 5.12

AEC note tags

Various types of tag can be used in AEC to label components, views and details. To access them pick [H18] on the tablet or pick Tags & labels ... from the Notes pull-down menu. The options available will then appear on the screen menu shown in Fig. 5.13.

Leader and text or balloon

The leader line function is one of the best in the Tags section giving you the option of a leader line of any length, either straight or kinked, followed by any text you may wish to add. The arrowhead is particularly effective and flexible in its selection and operation.

To point out and annotate a detail, as done in many of the illustrations in this book:

1   Pick [G18] from the tablet or Tags & labels ... from the Notes pull-down menu as shown in Fig. 5.7.
2   Pick Leader & text from the screen menu as shown in Fig. 5.13.
3   Set up the arrow type and size, text height, layer and the Paper space toggle to your choice and click on OK as shown in Fig. 5.14.
4   Prompt: Leader start: pick a start point on screen.
5   Prompt: To point: drag the rubber band to point where you wish the leader to change direction or the text to start.
6   Prompt: To point: leader can now be dragged in another direction (or press *enter* to begin typing text).
7   Prompt: To point: *enter* if dragged to another position.

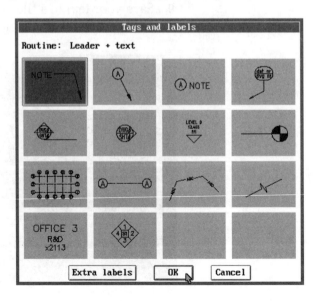

Fig. 5.13

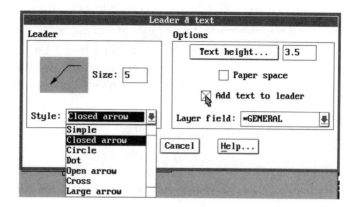

Fig. 5.14

Leader arrow will now appear on screen as shown in Fig. 5.15.

8   Prompt: Enter text: type the required text in the text entry
    dialogue box that appears on screen.
9   Prompt: Insertion point: place on screen where desired, as
    shown in Fig. 5.15.

Alpha note

This option places a bubble with an alphanumeric label on screen
followed by any text, as also shown in Fig. 5.15, using the same
routine as for the leader and text or balloon above.

Elevations

This routine allows the automatic underlining of text used to label
plans and elevations and also the insertion of an elevation number.

Break lines

This symbol can be used to indicate breaks in sections. Prompts ask
for insertion position, then you drag the symbol to enlarge it and
finally rotate to any angle.

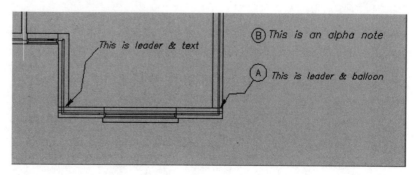

Fig. 5.15

Labelling rooms and adding attribute handles

Used to label a room with alphanumeric tags. Prompts ask you for room name which can be picked from a screen menu or can be typed in at the command line. You will also be prompted for an insertion point so that the symbol can be placed anywhere on plan.

Structural grid

Used to annotate a structural grid on a plan alphanumerically or with any system of your choice. This function is fully detailed in the AEC manual so its use will not be covered here.

Title symbols

Various title symbols can be picked from [H19] on the tablet or from the Notes pull-down menu. These include North arrow, Revision number and Revision cloud, among others.

CHAPTER 6

Third project: multi-storey domestic construction

The aim of this project is to serve as an introduction to modelling a multi-storey building by moving, copying and editing previously drawn elements of the single-storey bungalow. This technique can be applied to any multi-storey project and it ensures that dimensional coordination is perfectly preserved from one floor to another. Individual floor plans can be taken off from the model and displayed in any way convenient to the designer at any stage in the design process. The construction elements that link storeys (levels) are of course lifts and staircases and many types can be modelled in 3D or drawn in 2D using AEC. As an introduction, a simple straight domestic staircase will be inserted to indicate the program's potential as a design tool in this area. Unlike the previous projects only an outline of the work needed to produce multi-storey buildings is included; it will allow the reader to make a start, however, using the routines listed. As indicated earlier in this work only layer, level and coordinate information are important enough to be included on the status bar along with the Ortho and Snap toggles. Using levels is a vital part of the complete AEC design process and up to 100 can be established, identified and individually modified using AEC. When you start a new project you are working on level 0 (the base level) which represents ground level and can have foundations drawn below it as illustrated in the first project.

6.1 Working with levels

To work on the first, second or any subsequent floor levels you need to establish (using the Add new levels ... command) a new level, within which to work, at a new base height set vertically above the base level.

Practical

During this session you will practise the following:

- Establishing a new level for the first floor.
- Copying and moving elements such as walls and windows to the first floor level.
- Making alterations to the first floor level.

Note To avoid repetition, the elements already created will be used for this project which also serves to show the flexibility of editing possible. On your taught CAD course you will probably be asked to design a two- or three-storey house at some time so you will probably start by designing a completely new ground floor plan to copy.

To establish the first floor level

1 Pick [P5–6] on the tablet or Level status ... from the Levels pull-down menu as shown in Fig. 6.1.
2 Pick Add new levels ... and set up one or more new levels as shown in Fig. 6.2.
3 Click OK to confirm as usual.

Note The height of level one is set at the wall top of the ground floor. Use the multiple level option to set up several floors with the same interval between each floor. If the floor spacing is irregular set up each floor separately, going back to the main dialogue box on each occasion.

The new level will now appear on the dialogue box as shown in Fig. 6.3; the current level is indicated by the asterisk on screen.

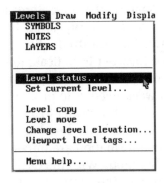

Fig. 6.1

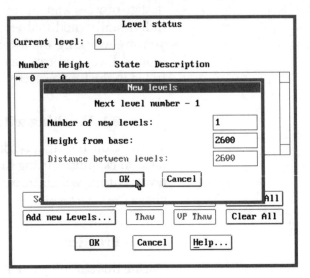

Fig. 6.2

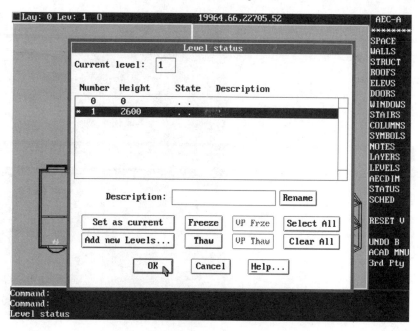

Fig. 6.3

Note   AEC always sets the current level back to level 0 after any level editing operation so always check the status line before starting work on the upper levels. The Set current level command is now obsolete as its functions are better done from the Level status dialogue box where all the other parameters can be checked at the same time. Levels can be frozen (made invisible and inaccessible) or thawed (visible and accessible) by simple clicks on this dialogue box and the results are covered in more detail below.

Copying the first level

Now copy the walls, windows and doors from the ground floor to the upper floor after first setting up two viewports (with the same viewpoints as for the bungalow model) so that the results of the copying can be seen.

1   Pick [P7] from the tablet or Level copy from the Levels pull-down menu.
2   Prompt: Select objects: **W** enter.
     Comment   It is easier to erase those objects not wanted later than to copy individual elements. The ground floor walls will have been blocked during the Wall convert command and may need 'Exploding'.
3   Window all the walls of the bungalow to select them.

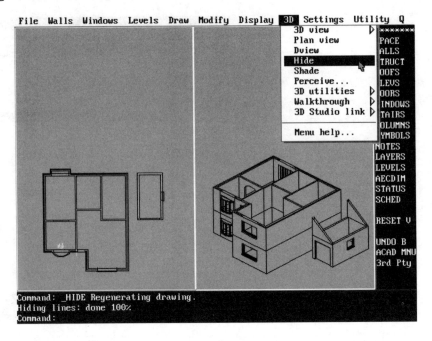

Fig. 6.4

4   Prompt: Select objects: *enter.*
5   Select the destination level or levels in the dialogue box that appears on screen.
    Comment   If level 1 was not previously set up in the Level status dialogue box an alert will appear reminding you to do so now.

The walls should now appear in both viewports but will only be seen within the right-hand port as shown in Fig. 6.4.

Prior to creating a multi-storey model save the bungalow using a different drawing name then remove the gable walls from one saved drawing using the AutoCAD Erase command and a window selection as shown in Fig. 6.5.

If you wish to use the same roof as the bungalow, freeze the roof layer for now, thaw it later and change the elevation to use it again.

The Level move command works in exactly the same way as Level copy except that a single image remains after the command has been used. Try moving one of the furniture units created earlier to the first floor.

Now you can start to work on the elements copied to level 1, but before starting any work use Level status ... to freeze level 0.

6   Pick Level status ... as before.
7   Set the current level to level 1, as shown in Fig. 6.6, by clicking Set as current and then clicking on level 1.

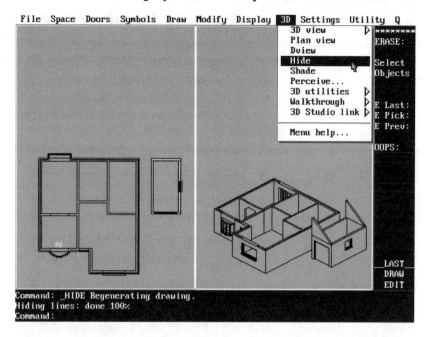

Fig. 6.5

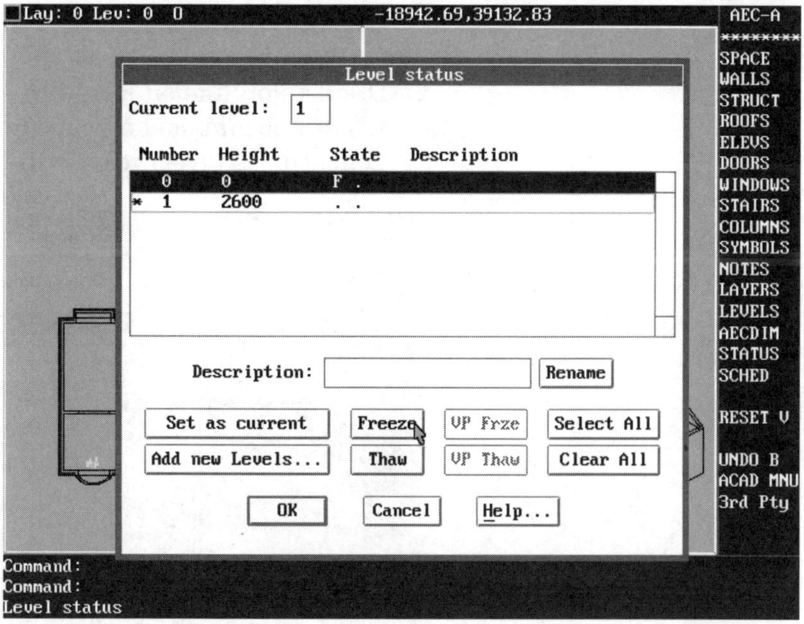

Fig. 6.6

Comment This will prevent any accidental alterations to the
ground floor (level 0) while you remove elements from the first
floor (level 1).

8   Click on level 0 to highlight it then click on Freeze as shown in
Fig. 6.6.

6.2 Inserting a flight of stairs

The following stage is a classic 'chicken and egg' design situation since you can insert a straight flight of stairs first and then establish a stairwell at the first floor level or vice versa. With CAD it will take only takes a few seconds to make any changes necessary later to either the stairs or the stairwell.

Practical

During this session you will practise the following:

• Insertion of a domestic staircase to link the two levels.

Inserting the stairwell

In this instance, the stairwell will be drawn first as a Polyline and the stairs inserted to align with the stairwell.

1   Pick [K10] on the tablet or AutoCAD commands, Polyline from the Draw pull-down menu.
     Comment   You should be familiar with this standard Auto-CAD command.
2   Draw a Polyline on level 1 to form a rectangle approximately 1000 × 3050 mm, as shown in Fig. 6.7.
     Comment   The staircase will be 900 mm wide. One partition

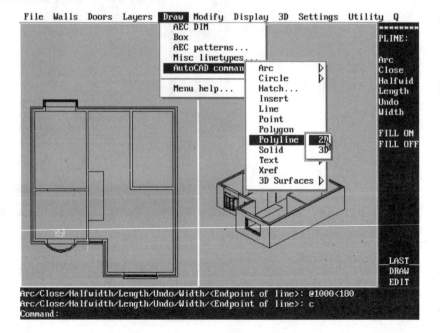

Fig. 6.7

wall that was copied from the ground floor has already been erased to make room for the stairwell.

Practice task

Now that the stairwell has been drawn re-plan the rest of the upper floor to create three bedrooms and a bathroom. Remove the copied windows then insert new windows to each room making sure that they conform to building regulation sizes. Use the Move command to move the stairwell towards the rear wall if this will allow more room for a bedroom in the front right-hand corner.

When the first floor plan is complete, and while level 0 is still frozen, use the Copy or Copy/Rotate commands to produce a copy of the first floor and place it on screen alongside the ground floor. You now have two separate floor plans both produced from the same model so that dimensional integrity is maintained.

Inserting the stairs

Now the staircase can be added either in 2D by picking [D11] on the tablet or 2D stair symbols from the Stairs pull-down menu or in 3D by a slightly more complex process. Since you are using the program for computer-aided *design* (not computer-aided *drafting*) the 3D symbol insertion will be used.

To insert the straight flight of domestic stairs you will attach it to the lounge/dining room partition wall as shown in Fig. 6.8.

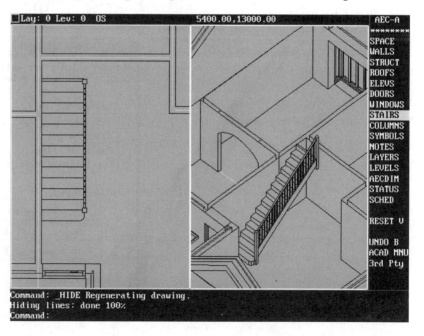

Fig. 6.8

Fig. 6.9

1 Pick [C10–11] from the tablet or Domestic staircase ... from the Stairs pull-down menu as shown in Fig. 6.9.

2 Set up the options in the Domestic staircase dialogue box as shown in Fig. 6.10 and click OK.

Comment   Pick the Menu help ... option on the Stairs pull-down menu for an explanation of terms in this dialogue box.

3 Select the Riser options in the dialogue box, as shown in Fig. 6.11, by clicking on one of the choices available.

4 Set up the Balustrading options as shown in Fig. 6.12 and click on OK.

Comment   Again check that the options selected comply with statutory requirements such as the Building Regulations. Note the complexity of the choices offered, which allows the development of a wide range of styles, and treat this routine as an introduction to the subject of balustrade design.

Fig. 6.10

```
┌──────────────────────────────────────┐
│          Domestic staircase          │
│ Alignment:              Left      ▼  │
│ 2D/3D Mode:             □ 2D ■ 3D    │
│ Riser height:           180    <<    │
│ Total rise:             2600   <<    │
│ Going length:           225    <<    │
│ Stair width:            900    <<    │
│ Handrail configuration: Right     ▼  │
│ Group staircase:        □ Yes ■ No   │
│      OK       Cancel       Help...   │
└──────────────────────────────────────┘
```

Fig. 6.11

```
┌──────────────────────────────────────┐
│              Riser options           │
│ The specified riser height will not fit
│ exactly within the given total rise. │
│                                      │
│    Current settings                  │
│    Total rise.................2600   │
│    Riser height..............180     │
│    Number of steps...........14      │
│    Gap to total rise height....80    │
│                                      │
│    Riser options                     │
│      13 risers at 200mm              │
│      14 risers at 185.71mm           │
│      15 risers at 173.33mm           │
│      Change total rise to 2520mm     │
│                                      │
│              Cancel                  │
└──────────────────────────────────────┘
```

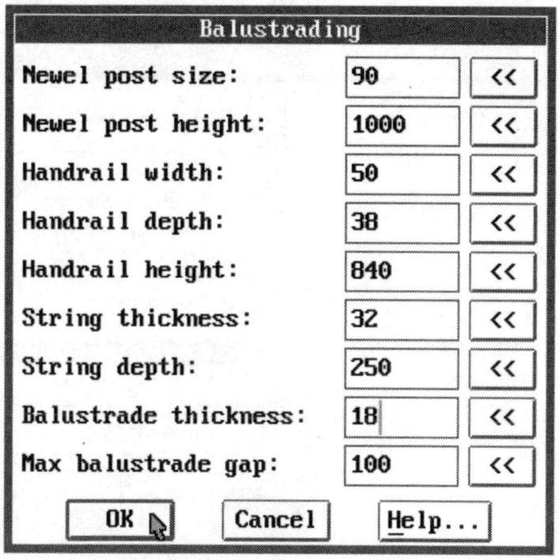

Fig. 6.12

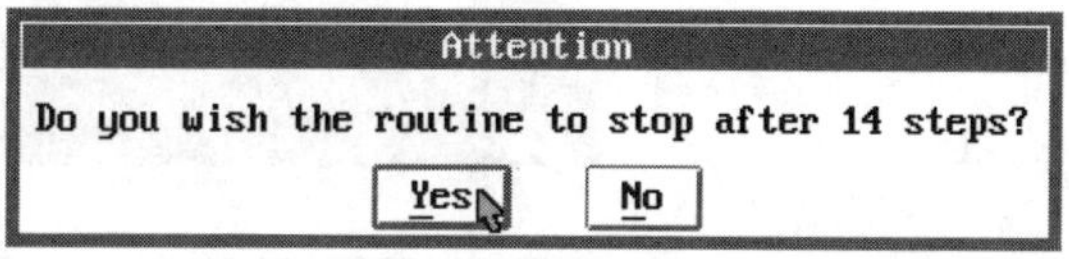

```
┌─────────────────────────────────────────┐
│             Balustrading                 │
│ Newel post size:      [90    ]    [<<]   │
│ Newel post height:    [1000  ]    [<<]   │
│ Handrail width:       [50    ]    [<<]   │
│ Handrail depth:       [38    ]    [<<]   │
│ Handrail height:      [840   ]    [<<]   │
│ String thickness:     [32    ]    [<<]   │
│ String depth:         [250   ]    [<<]   │
│ Balustrade thickness: [18    ]    [<<]   │
│ Max balustrade gap:   [100   ]    [<<]   │
│       [  OK  ]   [ Cancel ]  [Help...]   │
└─────────────────────────────────────────┘
```

Fig. 6.13

```
┌──────────────────────────────────────────────┐
│                  Attention                     │
│ Do you wish the routine to stop after 14 steps?│
│              [ Yes ]      [ No ]               │
└──────────────────────────────────────────────┘
```

5   Click on Yes in the dialogue box shown in Fig. 6.13.

6   Prompt: Enter staircase base point: use OSNAP NEA to pick any point on the partition wall as shown in Fig. 6.8.

7   Prompt: Enter starting direction <0>: **90** *enter*.

8   Click Yes in the Attention dialogue box to start insertion with a bullnose step.

9   Prompt: Quit/Landing/Nosing/Undo/Winders/<Number   of steps>: **12** *enter*.

   Comment   Nosing counts as one step in the step count. Note the complexity of options available to configure a domestic stair.

10 Prompt: Enter nosing length <100>: **50** *enter*.

6.3 Inserting the first floor

During this session you will practise the following:

•   Insertion of a hollow timber first floor.

Practical

Finally, insert the first floor around the stairwell.

1   Pick [E4] from the tablet or Construct floor ... from the Roofs pull-down menu as shown in Fig. 6.14.

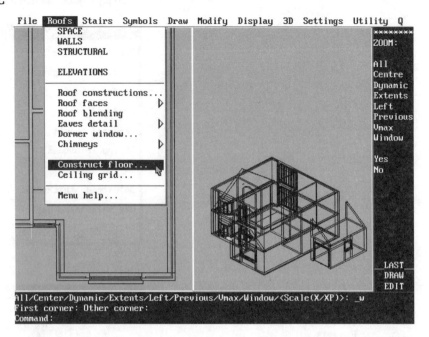

Fig. 6.14

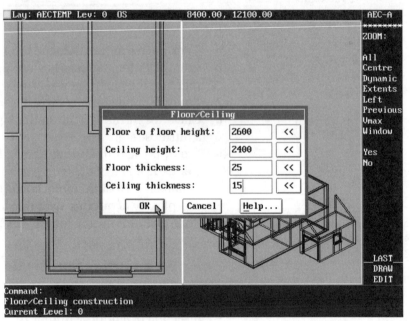

Fig. 6.15

2   Select the options shown in Fig. 6.15 from the Floor/Ceiling dialogue box and click OK.

Comment   Both the floor and ceiling thickness can be set up in this 3D routine; service runs and ducting can be inserted in 3D later.

3   Pick points to locate the floor.

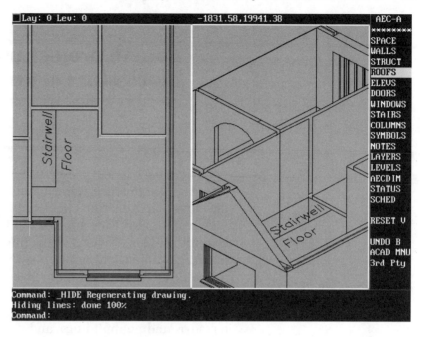

Fig. 6.16

4   Pick the corners of the room and stairwell to establish the floor edges as shown in Fig. 6.16.

Comment   Select your points all clockwise or all anti-clockwise.

Practice Task

Now insert the rest of the first floor – including walls, windows and doors – to your own design. Use the Change elevation [G12–13] command to reposition the existing roof and complete the multi-storey model.

Note   In the next chapter you will learn how to manipulate the 3D models created to produce a variety of different designs quickly and easily.

CHAPTER 7

Fourth project: making and using blocks for estate development

The standard practice, using manual paper technology, of an estate developer, architect or client is to submit plan and elevational views of a project for approval to the local authority's planning committee. This provides visual information only in 2D form with height (Z-dimension) information written on the plan at regular intervals or drawn as contour lines of equal height. This method provides the typical lay person with generally insufficient information to make an informed decision on the effect of the proposal on surrounding buildings and landscape. The results of past procedures are apparent everywhere in the built environment, especially on British housing estates, surely some of the worst planned and most ugly in Europe.

Many European countries, particularly Alpine countries, have tried to improve matters by insisting that profiles showing the vertical height (Z-dimension) of the proposed building be erected on the development site to check the visual impact, or insisting upon the presentation of 3D scale models.

Three-dimensional computer-generated models of large sites appeared a few years ago. These needed expensive specialist equipment, the cost of which prohibited their use outside all but a handful of large organisations.

Now 3D modelling can be achieved on a PC in the smallest practice, with modest capital expenditure. With the early PCs, however, much patience was required, even if you had the most powerful machine available for this work. In 1988, a 25MHz 386-class PC with a graphic's accelerator could produce 2D CAD drawings with reasonable speed. Modelling a small estate of, say, ten houses plus shops, garages, roads, trees and other features took one weekend to generate and the following weekend to plot but you did get a true, realistic 3D view of the proposed site for a fraction of the cost of making a real model.

Now, with more powerful PCs, the same job can probably be done overnight including plotting. It is sometimes easy to forget that with computer-based systems people are often not needed after the input stage and most output systems can do their work just as efficiently while there is no-one on the premises. With modern inkjet or thermal roll-feed plotters many plots can be made unattended overnight.

AutoCAD AEC's ability to create blocks (grouped drawing entities) of several types which can then be manipulated in many ways provides a very powerful 2D or 3D tool for estate developers planning a site for housing or industrial units. The general structure and operation of AutoCAD blocks has been covered in many previous works, so this work covers the application of AEC blocks for building and civil engineering use only.

Blocks in 2D or 3D, which can be made up of any standard components, may be floor plans, house styles, bungalows, garages, industrial units, shops, offices, etc. Once made these can be inserted into the current drawing many times but not into other drawings. Howvere, blocks can be written as a .dwg file to the hard disk and then inserted into any other named drawing one at a time or in multiple units onto the site plan in 2D or into the site model in 3D.

The important question is whether blocks should be made in 2D or 3D.

For true site modelling 3D blocking is essential, but you may only need to use 2D blocks in layout, location or site plans for legal and/or planning purposes at the moment. (Remember though that proposed US and EEC legal requirements will require 3D planning submissions within the near future.)

Users of earlier versions of AEC recognised that more flexibility with 2D and 3D insertions was required and with Version 4 you can use both methods. Conversion of entities from 3D to 2D for block generation is covered in the practical routines that follow.

Working through the following practical section is the only way to fully appreciate the power of blocks and the flexibility that they give to practising designers in everyday use. In general terms, once the form of any item more complex than a single line or curve has been established it should not be redrawn again but become part of a block.

Comprehensive features exist in AutoCAD AEC for editing established blocks, which can be used to easily create many design variations on a standard theme; this is a technique which artists and composers have used in their work for centuries.

Note   Use a package such as Softdesk's Digital Terrain Modeller (DTM) to produce a ground model of the site with pre-prepared foundation pads for any buildings shown at the correct level.

7.1 General block commands

Block

Pick [P8] on the tablet or type in the command at the keyboard. Used to group together drawing entities contained in the project models for insertion into another part of the same drawing file.

Wblock

Pick [Q8] on the tablet or type in at the keyboard. Used to write a block to the hard disk (a 'written block') for storage and future insertion into another drawing at any X, Y and Z scale desired and at any level or elevation.

Xref

Pick [P9] on the tablet, from the Draw pull-down menu or type in at the keyboard. This is a special type of block that enables the insertion of entities or of complete drawings that are frozen in content and cannot be changed within the current drawing editor. This command is very valuable where a team of designers all work on the same reference model. The architect's floor layout could be worked on by the services engineer who might be tempted to push the walls outwards to accommodate extra heaters. If the original floor plan is attached as an Xref, it cannot be changed by the services engineer who has no access to the original drawing. In the site design section that follows, the attachment and binding of Xrefs is particularly useful.

Insert

Pick [R8] on the tablet, from the Draw pull-down menu or type in from the keyboard. Used to insert AEC blocks, written blocks and your own pre-defined blocks into the same drawing or other drawings. Also used in a similar way for insertion of RIBACAD or other symbol library details. Probably the most useful and most used command for the experienced AEC user.

Minsert

This command can only be typed in from the keyboard without switching to the main AutoCAD menu. The command enables the insertion of multiple copies of blocks into another drawing and makes for speedy insertion of repetitive designs, such as terrace housing, but which restricts the editing possibilities for individual blocks.

7.2 Block creation

First make a copy of the model to work on so that the original cannot be corrupted during the routines that follow. This may entail using the Setup command to enlarge the drawing sheet size again.

Practical

During this session you will practise the following:

- Copying the master model and work on the copy when blocking and write blocking.
- Separation of the various floor plans or entities that require to be created as blocks.
- The use of the 2D/3D convert feature to produce suitable views for blocking.
- Creation of 2D and 3D blocks of standard dwelling types.

Making 3D blocks

To make a block of the showroom, bungalow or house model for insertion into another part of the same drawing:

1 Pick [P8] from the tablet menu or type in Block at the keyboard.
2 Prompt: Block name (or ?): **BUNG1** enter (see Fig. 7.1).
   Comment   The '?' option gives list of blocks already defined.
3 Prompt: Insertion base point: pick OSNAP END on an exterior wall as shown in Fig. 7.2.
   Comment   This must be a point at the base of the walls in elevation.

Note   The insertion base point should be chosen with care, particularly if you are blocking in 3D. If any point above elevation 0 is picked, such as a gable wall endpoint, the block will not be inserted at the correct level.

4 Prompt: Select objects: **W** *enter.*
   Comment   Any view of the model can be made into a block for insertion or just a view of part of the model; blocks can also be nested.

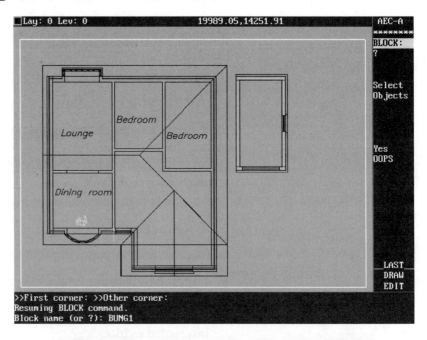

Fig. 7.1

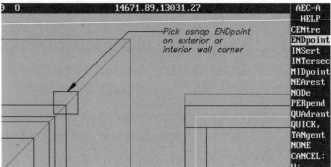

Fig. 7.2

5   Prompt: First corner: Other corner: pick two corners that
     include all the drawing entities to be blocked.
     Comment   The number of entities found will we displayed.

The entities found will display as dotted lines as shown Fig. 7.3.

6   Prompt: Select objects: *enter.*
7   Items blocked will now disappear from the screen.
     Comment   Use the Oops command from the screen menu to
     bring them back on screen if required or use the Insert
     command to place them in another position. The bungalow and
     garage could have been blocked separately for inclusion in
     different site positions in the next chapter.
8   Pick [W18] from the tablet or Oops from the screen menu if you
     want the blocked entities back on screen as shown in Fig. 7.4.

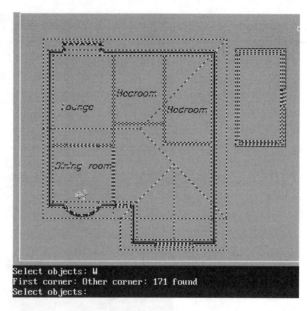

Fig. 7.3

Making 2D blocks

First convert the copy of the model from 3D to 2D using the following routine.

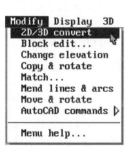

Fig. 7.4

1 Pick [H12–13] on the tablet or 2D/3D convert from the Modify pull-down menu as shown in Fig. 7.4.

2 Pick the Conversion mode from the dialogue box that appears on screen as shown in Fig. 7.5.
  Comment Entities converted will be specified as will those filtered out.

3 Erase the roof and any other entities that were filtered out during the conversion process.

4 Prompt: Select objects: pick Window from the screen menu and window the model.
  Comment Note how the display of windows and doors has changed. Check that all conversions have taken place by viewing in 3D. Walls are not converted but remain at the height declared in 3D as this does not affect 2D plotting.

5 Prompt: Select objects: *enter.*

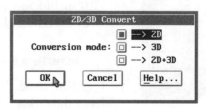

Fig. 7.5

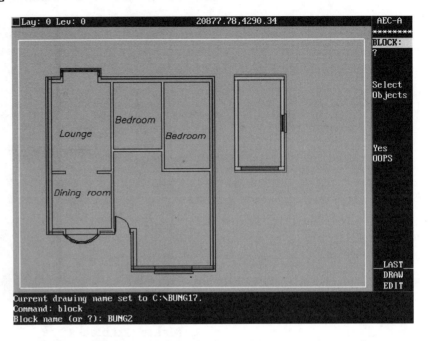

Fig. 7.6

6   The conversion process will produce entities as shown on screen in Fig. 7.6.
7   Make a 2D block of the copy, in the same way as for the 3D block, using the block name of BUNG2 and being careful to pick an insertion point at ground level away from the gable walls.
8   Leave the screen blank as shown in Fig. 7.7.

7.3 Block insertion and editing

Use the Insert command to insert the blocks created into the same drawing. If the blocks are to be inserted on any level other then level 0 remember to use the Level status ... Level set commands to ensure that the blocks are inserted on the correct level.

Practical

During this session you will practise the following:

*   Insertion of blocks into the same drawing.
*   Editing blocks so that they can be used for different purposes.
*   Use of written blocks to insert blocks into any drawing.

Inserting blocks

Fig. 7.7

1   Pick [R8] from the tablet menu or AutoCAD commands, Insert from the Draw pull-down menu as shown in Fig. 7.8.
2   Click on Block in the Insert dialogue box, as shown in Fig. 7.9.

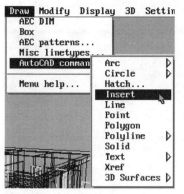

Fig. 7.8

3 Pick BUNG1 as the block name to insert from the selection of defined blocks, as shown in Fig. 7.9.

4 Click on Specify parameters on screen and pick Explode if you want the block to be exploded when inserted (use settings as shown in Fig. 7.10).

Comments Only select the Exploded option if your sure that modifications will be made to the block after insertion.

5 Prompt: Insertion point: pick any point on screen.

6 Prompt: X scale factor <1>/corner XYZ: **0.5** enter.

Comment As the original was drawn at 1:50 this gives a scale of 1:100 for the inserted block. Insertion at any of the normal scales can be done by decimal reduction.

7 Prompt: Y scale factor <default=X>: *enter*.

Comment You could change the proportions of X to Y scaling at this point.

8 Prompt: Rotation angle <0>: **45deg** enter.

Comment A random rotation angle has been selected to illustrate the function.

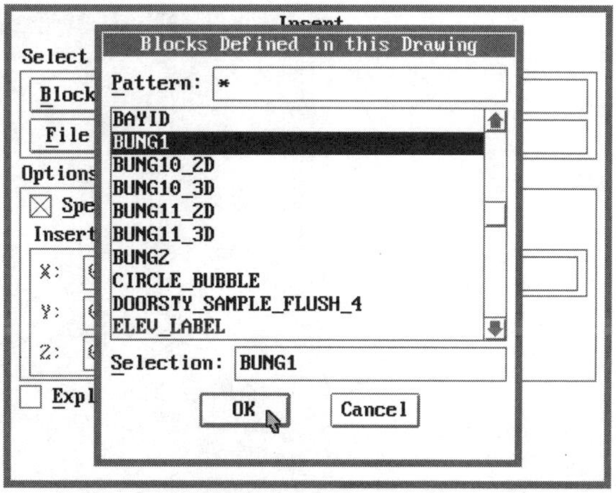

Fig. 7.9

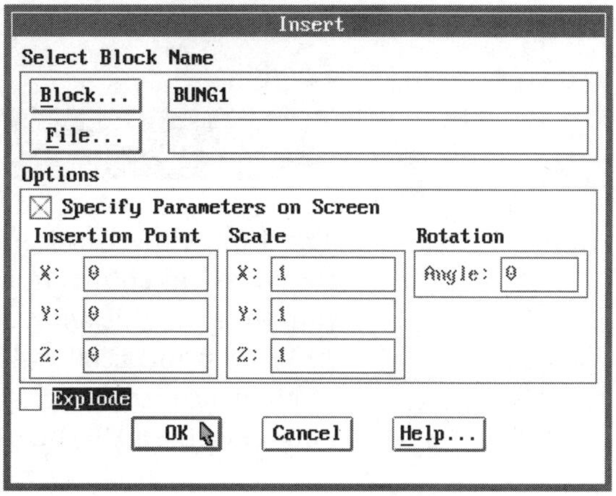

Fig. 7.10

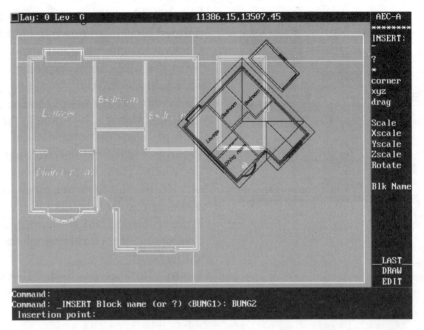

Fig. 7.11

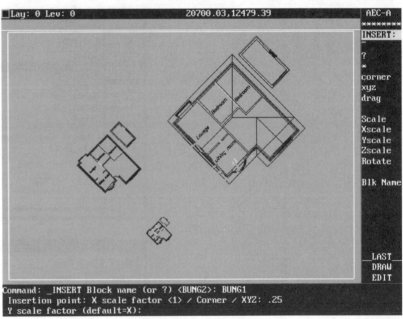

Fig. 7.12

The bungalow block should now appear on screen, at the scale requested, as shown in Fig. 7.11. Next insert the 2D block BUNG2 in the same way, as shown full size ghosted (in Fig. 7.11) and solid at 1:100, 1:200 and 1:500 scales (in Fig. 7.12).

Now use any AEC 3D viewing feature to see the difference between 2D and 3D blocks on screen.

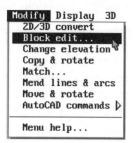

Fig. 7.13

Editing blocks

AutoCAD AEC has a comprehensive block editing facility for the rapid alteration of several features connected with scales and rotation of inserted blocks. These will need to be used when building up a site plan in the next chapter so we will examine the possibilities here.

1   Pick [A12–13] on the tablet or Block edit … from the Modify pull-down menu as shown in Fig. 7.13.
2   Prompt: Select block: pick any point on the block that you need to edit.
3   Pick the value boxes required from the block edit dialogue box, type in the new values then click on OK as shown in Fig. 7.14.
    Comment   Select any different X, Y and Z scale values and rotation angle. Do not forget that you can use the Block edit … command to alter not only your own blocks but also any RIBACAD or pre-drawn AEC blocks.
4   View the result in 2D or 3D on screen.

Inserting blocks of models into other drawings

Once standard house types or industrial units have been drawn as a 2D plan or 3D model and enhanced with surface textures, they can be inserted on site or layout drawings at a different scale to form a 2D plan or 3D site model which consists also of roads and landscapes.

   If you are creating a 3D model using the Level status … and Level set commands you are able to set buildings on site with each one at a different finished floor level. A software package, such as Softdesk's Digital Terrain Modeller, can then be used to design roads, paths, sewers, gardens, retaining walls and other civil engineering works. Site modelling in 3D is outside the scope of an

Fig. 7.14

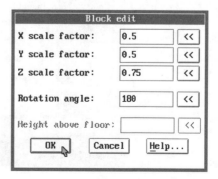

introductory work so the site used for this and the next chapter is assumed to be flat.

Writing blocks to disk as .dwg files

The various blocks, whether 2D or 3D, need to be stored on your hard disk under file names that can be accurately catalogued. This routine is of vital importance to most AEC users as it allows complete drawings that show part of a project to be inserted into other drawings at the same or different scales.

To write block BUNG1 to the hard disk for insertion in any other drawing:

1   Pick [Q8] from the tablet or type in Wblock at the keyboard.
2   Click on the Type it option in the Create drawing file dialogue box and type in BUNG25 as the file name.
     Comment   You can use the same name for the block and written block.
3   Prompt: Block name: **BUNG1** enter.
     Comment   BUNG1 is now a drawing file written to the hard disk as BUNG25.DWG for future use. As a drawing file it can be Inserted or Minserted into any other drawing in the same way that any drawing file saved to disk may be inserted by using the Insert command. Figures 7.15 and 7.16 shows the file format created and typical insertion parameters for the block to be inserted at a scale of 1:500.

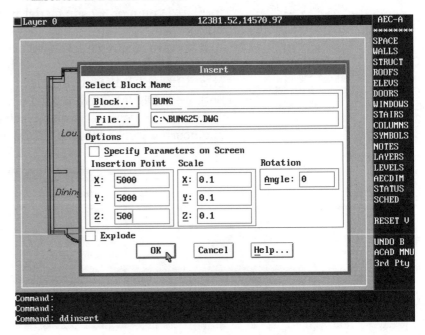

Fig. 7.15

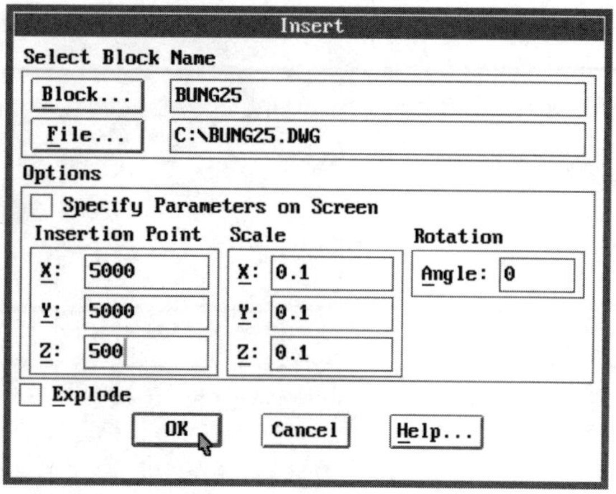

```
┌─────────────────────────────────────────────────┐
│                     Insert                        │
├─────────────────────────────────────────────────┤
│  Select Block Name                                │
│                                                   │
│   ┌─────────┐   ┌─────────────────────────────┐  │
│   │ Block...│   │ BUNG25                      │  │
│   └─────────┘   └─────────────────────────────┘  │
│   ┌─────────┐   ┌─────────────────────────────┐  │
│   │ File... │   │ C:\BUNG25.DWG               │  │
│   └─────────┘   └─────────────────────────────┘  │
│  Options                                          │
│   ┌──────────────────────────────────────────┐   │
│   │ ☐ Specify Parameters on Screen            │   │
│   │  Insertion Point   Scale        Rotation  │   │
│   │  X:  ┌──────┐    X: ┌─────┐   Angle: ┌──┐ │   │
│   │      │ 5000 │       │ 0.1 │          │ 0│ │   │
│   │      └──────┘       └─────┘          └──┘ │   │
│   │  Y:  ┌──────┐    Y: ┌─────┐               │   │
│   │      │ 5000 │       │ 0.1 │               │   │
│   │      └──────┘       └─────┘               │   │
│   │  Z:  ┌──────┐    Z: ┌─────┐               │   │
│   │      │ 500  │       │ 0.1 │               │   │
│   │      └──────┘       └─────┘               │   │
│   └──────────────────────────────────────────┘   │
│   ☐ Explode                                       │
│        ┌──────┐    ┌────────┐   ┌────────┐        │
│        │  OK  │    │ Cancel │   │ Help...│        │
│        └──────┘    └────────┘   └────────┘        │
└─────────────────────────────────────────────────┘
```

Fig. 7.16

Practice task

Write block BUNG2 to the hard disk using a name of your own choice.

Note   The next logical stage will be a legal requirement for all developers to submit planning applications not on paper but as a real-time multimedia presentation (produced on a CD, or written directly from a rendering package to screen, with hard copy reference material). In the USA and Scandinavia it is also becoming common practice to issue contract documents including drawings on CD to reduce the paper mountain involved with traditional tender procedures.

AutoCAD now has the ability to nest finished drawings in the current drawing editor which retain the features of the existing drawing and are frozen entities. Many details created by different design disciplines can be brought together in this way for plotting convenience. The Xref command gives access to this section of the program and is picked from [P9] on the tablet or from the Draw, AutoCAD commands pull-down menu.

Unfortunately use of the Xref command adds another level of complication to the blocking process which is outside the scope of this introductory volume. However, Fig. 7.17 shows file BUNG25 loaded as an Xref in the current drawing editor.

Block and Wblock insertions

The 3D or 2D bungalow block can now be inserted anywhere on the existing drawing, at any X, Y and Z scale factors. The written block version of the block can be inserted into any other drawing

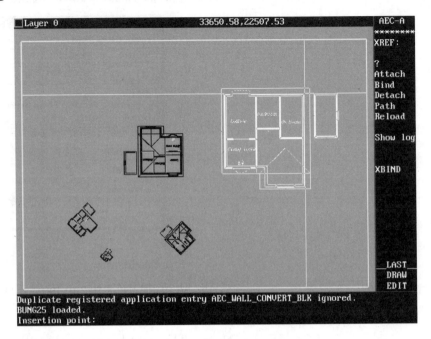

Fig. 7.17

using the Insert or Minsert commands to build up estate layouts and finished plans or 3D models.

Practice task

Make written blocks of the house and showroom models in 2D and 3D for later insertion onto a site plan. This is also a good time to make blocks of the individual floor plans from the 3D house model.

Use the Level display command to freeze all the levels except the one level that you want to copy. Then use Copy/Move to show each floor on a different part of the drawing sheet as described in the previous chapter.

7.4 Creating a site plan

Now you can use the blocks and written blocks made in the previous chapter as a very time efficient method of planning a small estate.

As this is the final project it is time to bring together clones of all the previous project models that you have created to produce a 2D site layout with 3D overtones. This will show part of a small estate made up from estate roads and pavements plus the following drawings:

- The garage or showroom drawing (SHOW1).
- The bungalow drawing (BUNG1).
- The house drawing (HOUSE1).

The rapid creation of site plans and estate layouts is so easy using AEC that its use should lead to more alternative designs being prepared more quickly than by existing hand-drawn methods. Availability of more choices should in turn lead to more intelligent decisions about where buildings are placed on available construction sites.

Site layout work can begin with the placement of buildings as suggested above or by first positioning the roads and footpaths then adding the proposed buildings. Once the building design and road planning is complete other features which are found on a typical site such as trees, shrubs, walls, cars and other forms of transport can be added as site symbols. Some symbols are 2D only, others are fully drawn in 3D.

As this is an introductory volume on AEC only the creation of 2D site plans is covered in detail although most of the elements inserted can be manipulated to form 3D symbols. Restrictions in the capability of the software are currently less significant than restrictions in the graphics capabilities of the hardware; however, computer hardware continues to develop rapidly. The new RISC-based 'PowerPC' machines will find a ready market in 3D graphics and CAD work which are ideal areas for this type of computing.

Most 3D features for site work could be modelled by the pioneers who used AEC Version 2 but the process was so slow that the results produced did not justify the effort expended.

Even with later releases it is important to prepare the 3D models by adding as much detail to the surfaces as possible before insertion to add realism to your designs. Unfortunately such detail, and the inclusion of internal elements, adds to Hide and Regen times so erase all the internal elements that are not required before blocking the models for site insertion. Practical notes on these stages are listed below.

Note   Try freezing layers as an alternative to erasing them.

Practical

During this session you will practise the following:

- The use of Block and Wblock with all the previous project models.

- Creating a new drawing with an A3 sheet size, at 1:500 scale and naming it.
- Drawing a simple road layout on screen which includes footpaths and kerbs.
  Comment   Consult the various design guides such as *Roads in Urban Areas* issued by the Department of Transport before deciding on features such as radii of curves, widths of estate roads and car parking dimensions.
- The use of the AutoCAD Edit commands to add features to the road layout.
- Assembling finished models on site (with Wblock), positioned anywhere for the initial insertion.
- The use of the Move/Rotate, Copy/Rotate commands to move the blocks into final position.
- Deciding upon suitable methods of providing site divisions.
  Comment   These can be shown as boundary walls, fences or other features simply by using AEC symbol insertions, creating your own 3D symbols to insert or using AutoCAD commands such as Single wall pick or just Polylines suitably edited to show plot boundaries.
- Adding site symbols to provide landscape detail and complete the plan.
  Comment   Many types of trees, cars, boats and other features can be added to complete the site. Pick all of them from the Symbols pull-down menu and then select the Site option or pick [H14–15] on the tablet menu. See the section below on inserting site symbols for more details.

Estate road detailing

To design and detail estate roads using the AEC Pavement command selected from the Site symbols menu:

1 Pick [H14–15] from the tablet menu or Pavement from the Symbols, Site pull-down menu as shown in Fig. 7.18.
2 Select the values shown from the Pavement dialogue box in Fig. 7.19.
   Comment   You have now set the parameters to draw multiple lines with one command in a continuous sequence and there must be no break in that sequence. Do not press return at any time until step 12 is completed.
3 Prompt: Draw kerb line BETween/RELative/<From point>: pick a central point on screen to start the kerb line.
4 Prompt: Arc/BETween/Length/RELative/Undo/<To   point>:

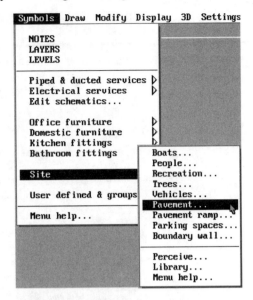

Fig. 7.18

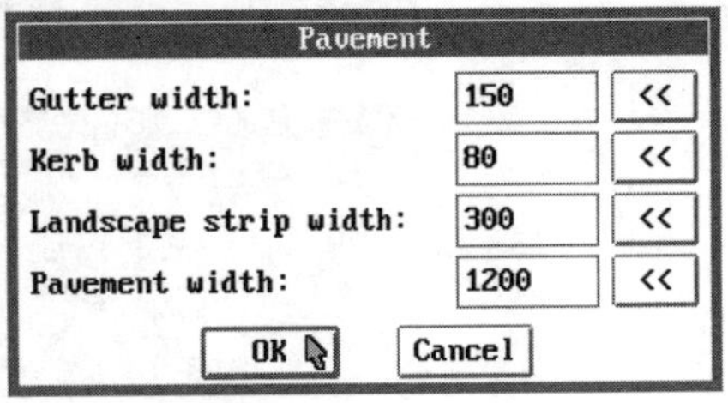

Fig. 7.19

With Ortho toggled on pull the rubber band 35,000 mm to the right and press the pick button.

5   Prompt: Arc/Length/RELative/Undo/<To point>: **A** enter.

6   Prompt: 90Deg/180Deg/Close/DIrection/DRag/Line/Second pt/Undo/<Endpoint of arc>: **90deg** enter.

7   Prompt: Arc radius: **10000** enter.

   Comment   Check your design guide regarding roads in urban areas to see if this is a suitable radius for a side turning leading from an estate road.

8   Prompt: Indicate arc direction: pick point above the horizontal line as shown in Fig. 7.20.

9   Prompt: 90Deg/180Deg/Close/DIrection/DRag/Line/Second pt/Undo/<Endpoint of arc>: **L** enter.

10 Prompt: Arc/Length/RELative/Undo/<To point>: drag rubber band upwards 25,000 mm with Ortho toggled on.

11 Prompt: Arc/Length/RELative/Undo/<To point>: enter to terminate sequence as shown in Fig. 7.21.

12 Prompt: Indicate pavement direction: pick point above the line on screen.

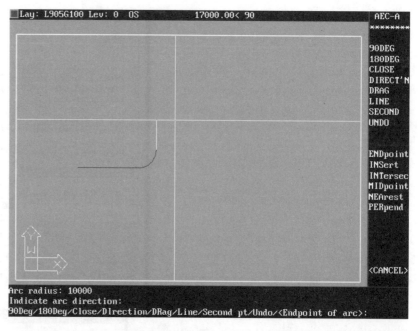

Fig. 7.20

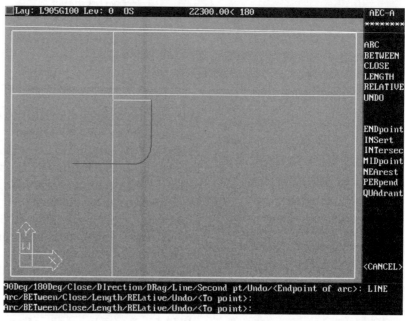

Fig. 7.21

Now Zoom window to see all the pavement elements, which should now appear on screen together as shown in 7.22. Note the automatic reappearance of the Site pull-down menu.

Note The example shown above only includes use of the 90Deg option in the formation of the estate road and path arcs. Other

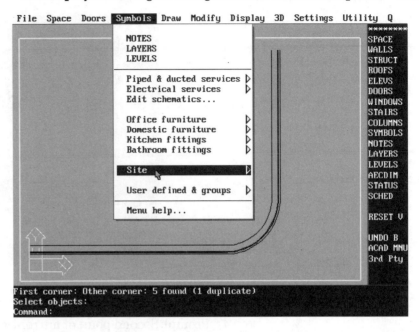

Fig. 7.22

options like Direction, Drag and Second point can be used to form estate roads with many different types of curve. A grid of construction lines possibly set up to standard road widths will be a useful aid to picking the points mentioned without the need to type in coordinates, angles and distances.

Other arc control methods include: the Direction option which is best started after the Line command has been used and with a known endpoint; and the drag option in which the line of the arc is dragged as the endpoint is moved further from the first point.

AutoCAD shortcuts

Both the half-road kerb and landscape strip can now be mirrored together to produce the other half of the road.

   To mirror the pavement and road edge created:

1   Pick [X12] from the tablet or Mirror from the Modify pull-down menu.
2   Prompt: Select objects: window all the entities on the completed side of the road.
3   Prompt: First point of mirror line: with OSNAP toggled on pick a point halfway across the road as shown in Fig. 7.23.
   Comment   Toggle Ortho on throughout these routines to ensure that kerb and mirror line are parallel and mirrored

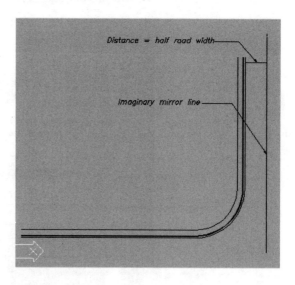

Fig. 7.23

parallel. Use a design guide to determine suitable standard road widths.

4   Prompt: Second point of mirror line: drag the shadowed entities on screen around the angles until they sit opposite the original entities then pick the point.

5   Prompt: Delete old objects? <N>: enter.

The completed half-junction should now appear on screen as shown in Fig. 7.24.

Note   The individual thicknesses of the kerb lines can now be changed using the Chprop command to represent a true kerb thickness for 3D work if required.

Now repeat the process to provide a complete road junction as shown in Fig. 7.25. AutoCAD commands such as Array, Offset, etc. are useful for site design.

Inserting previous project drawings

Now the other drawings created earlier in the book can be inserted on the site plan and then moved and rotated into final position.

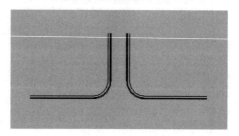

Fig. 7.24

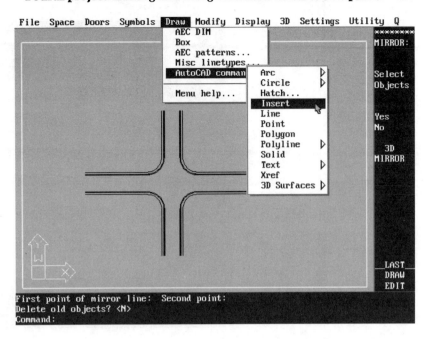

Fig. 7.25

1   Pick [R8] from the tablet or Insert from the Draw, AutoCAD commands pull-down menu as shown in Fig. 7.25.
2   Click on File in the Insert dialogue box.
3   Pick SHOW1 from the Select drawing file dialogue box as shown in Fig. 7.26 (or type the filename).

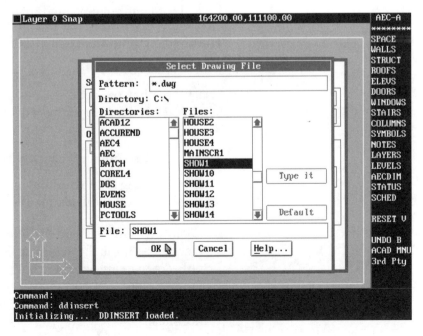

Fig. 7.26

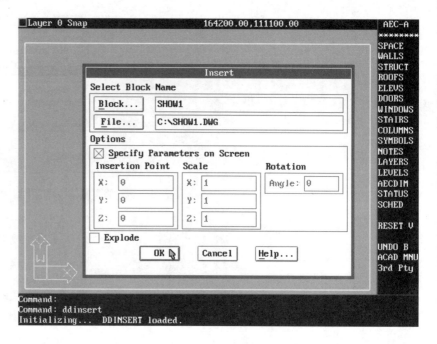

Fig. 7.27

Comment   For clarity in the screen shot the drawing file is shown in the root directory. Normally all drawings would be saved into a separate sub-directory.

4   Click on OK in the Insert dialogue box as shown in Fig. 7.27.
5   Prompt: Insertion point: pick a point on screen and confirm a scale of 1 and rotation angle of 0.

The screen will now appear as shown in Fig. 7.28.

Practice task

Insert the other project drawings in the same way with each one being placed on a different part of the screen.

Manipulation of inserted blocks

Final positioning of the building plans on site is achieved by using the combined AEC Edit commands Move/Rotate and Copy/Rotate.

1   Zoom a window around the final positioning area.
2   Pick [C12–13] on the tablet or Move/Rotate from the Modify pull-down menu as shown in Fig. 7.28.
3   Prompt: Select objects:
    Prompt: Base point or displacement: pick any suitable point on the block.

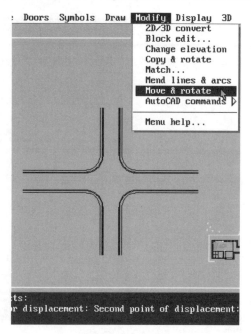

Fig. 7.28

4   Prompt: Second point or displacement: pick a point on the site
    to position the block.
5   Prompt: <Rotation angle>/Reference: drag the block into posi-
    tion or type in a suitable reference angle.

The block will now appear on screen as shown in Fig. 7.29.

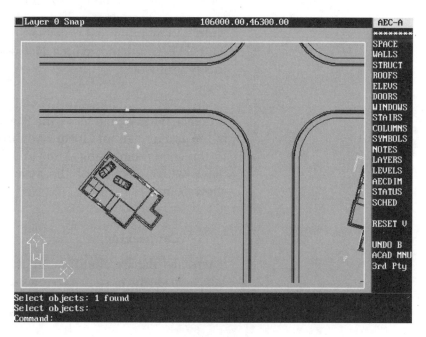

Fig. 7.29

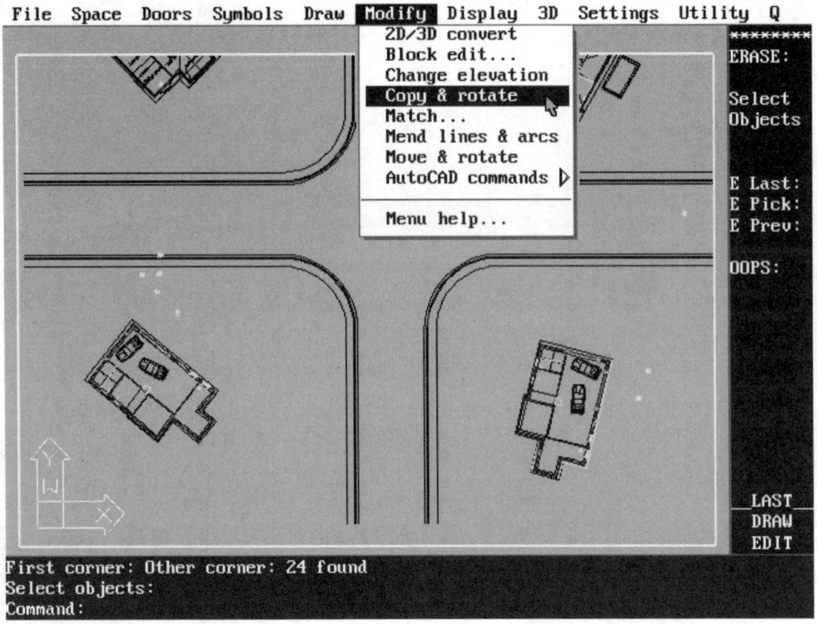

Fig. 7.30

Now the Copy/Rotate command can be used to produce further
insertions of the same drawing or written blocked drawing as
shown in Fig. 7.30.

The various house and bungalow blocks that have been written
to disk earlier (using the Wblock command) can now be added to
the site model using the insert command.

Note   Remember that pressing the space bar repeats the previous
AutoCAD command, which is a useful time-saving shortcut for
this type of work.

Practice task

Enhance the site plan using the Pavement command to extend one
of the estate roads and form a hammer head. Then provide a drive
to the showroom using AutoCAD Edit commands to merge the
drive with a section of the pavement to provide access to the
showroom.

Car parking

There are also excellent facilities within AEC for creating other
landscape features such as boundary walls, car parking spaces, car
ramps and vehicles.

Some of these are included in site symbol insertions below but others require more advanced skills than can be covered by a brief description.

Browse the library slide files found by picking library from the Site symbols pull-down menu as shown already in Fig. 7.18.

Note   You will probably realise by working through this project how often the Insert command will be used in everyday CAD work. When combined with the various block commands an environment for the very rapid production of design drawings is created.

It seems strange then that the software designers at Autodesk make the user search through inconsistent pull-down menus combined with keyboard input to access these commands. Only on the tablet menu are they sensible grouped in logical order, although even here some have been omitted. They should be made accessible through on-screen icons or special shortcut keystrokes.

Site symbol insertion

To complete this project landscape features can be added to finish the plan or 3D model. Many hundreds of pre-drawn 2D and 3D symbols can be inserted to enhance site plans, some of which are illustrated below. Many are parametric in nature giving several thousand more variations for the designer to use.

You can choose which symbols to use on your site drawing; use the insertion process which has been fully described previously. The following figures show some of the variations available.

Vehicles in 2D and 3D

3D and 2D cars, buses and goods wagons are available (Fig. 7.31 and 7.32).

Trees and plants in 2D and 3D

Trees and plants can be inserted with different heights and widths (Fig. 7.33 and 7.34).

Recreational facilities in 2D

Various types of sports facility on plan including badminton, football, pools, tennis and an athletics track are available (Fig. 7.35).

Garden furniture in 2D and 3D

Figures 7.36 and 7.37 show examples of garden furniture in two- and three-dimensions.

Garden furniture can also be used to complete the picture. Use

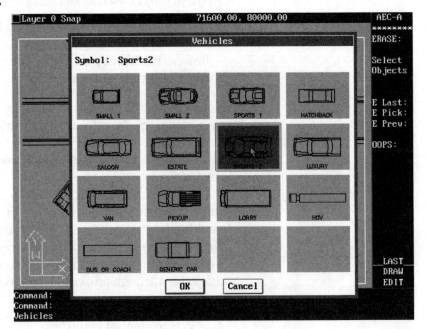

Fig. 7.31

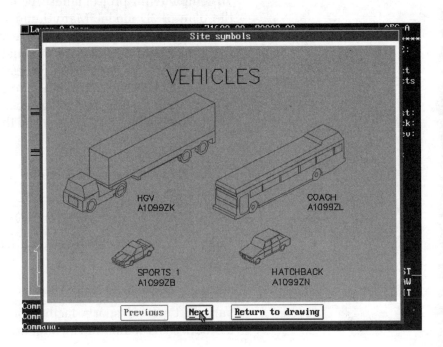

Fig. 7.32

Fig. 7.33

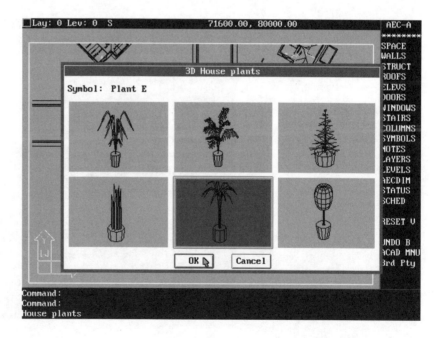

Fig. 7.34

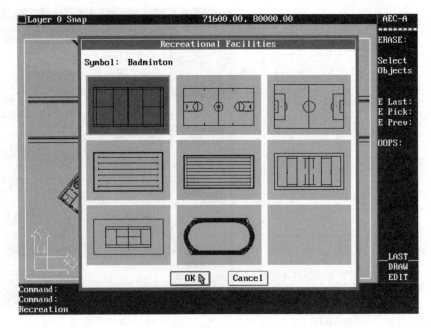

Fig. 7.35

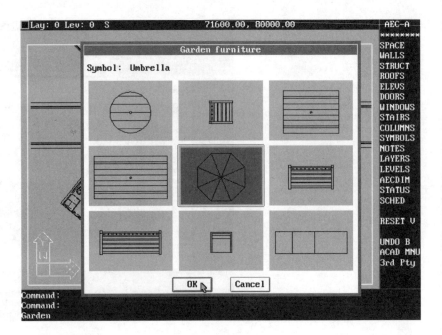

Fig. 7.36

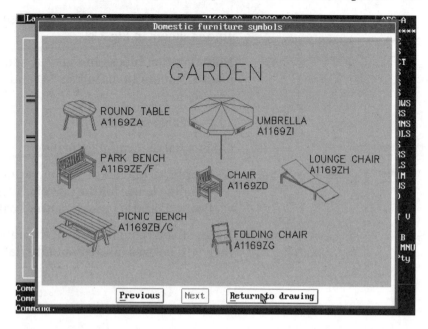

Fig. 7.37

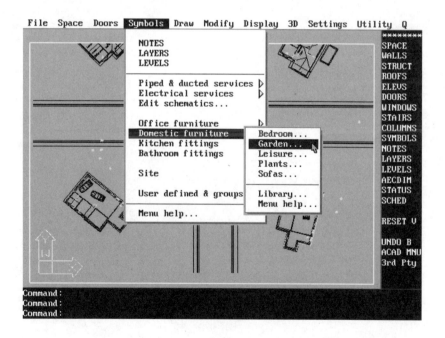

Fig. 7.38

the domestic furniture library to see the garden furniture available as shown in Fig. 7.38.

Note Written blocks in Figs 7.32 and 7.37 have associated written block numbers; items in Figs 7.31 and 7.33–36 can be picked directly on screen.

Other symbols

Other symbols available include:

- People: 2D and 3D people in various positions like walking and standing to use for the perceive command.
- Boats: 2D and 3D yachts, power boats, rowing boat and canoe.

Note   Even though many of the above symbols are in 3D the site they will be used on is essentially flat in this project.

CHAPTER 8

Printing and plotting

The AutoCAD AEC Plot command on the File pull-down menu or
the tablet [W24–25] supports a vast range of laser printers and
inkjet printer/plotters together with almost every make of pen
plotter. (For details see the Reconfiguring AutoCAD section p.000).

Other methods of producing hardcopy via alternative routes
exist within the Import/export option also on the File pull-down
menu. Encapsulated Postscript (EPS) files can be written for output
to any device that supports the Postscript language or for importation into other graphics applications. Drawing files in DXF format
can also be created and exported to other packages.

If the AutoCAD 12 Render option is used then TIFF and TGA
files can be created and exported to image processing applications
for the enhancement and production of many types of photo-
realistic and artistic colour output.

Hardcopy devices

Until the very recent past the standard method of output for large
format drawings has been the pen plotter. Average CAD drawings
have rapidly become more complex and ever larger in size and the
venerable pen plotter has been unable to keep up with the amount
of output required by the average design office. When you consider
that the electronic file size for the floor plan for a large building
may be 5 Mb and that the pen plotter could take nearly four hours
to plot such a file then it becomes almost impossible to justify the
time taken, even though the plot quality will still be better than
most of the other available technologies.

These drawbacks to the pen plotter have led to the development
of large-format inkjet printers which, although at first were
expensive due to the small quantities produced, have rapidly come
down in price.

Large-format mono inkjet printers with 600 dots per inch (dpi)
output will plot a 5 Mb file in ten minutes with only a slight
reduction in output quality. Have a good look at the large-format

colour inkjet printers available which are getting cheaper and better specified all the time.

8.1 Printer and plotter types

These are discussed in order of current market popularity but the reader should be aware that plotter technology has changed at a bewildering rate in recent years and still appears neither mature or settled.

Inkjet and bubble jet

Within just two years during the early 1990s this device became the standard method of outputting electronic large-format drawings. At first they were only available as mono devices but are now also full colour, or switchable between mono and full colour using interchangeable cartridges. Several variations on the basic technology are competing for a rapidly growing market with falling price levels making further improvements viable. They are very simple to set up and use but they lack the versatility of the pen plotter. The much higher plot speed more than makes up for the lack of choice for the output media and other factors such as the high cost of consumables compared to a pen plotter.

Laser

Laser printers were once only available as mono A4 devices but are rapidly getting bigger engines that can cope with A3, and even A2, output. Resolution technology is allowing production of 600 and 1200 dpi output at very low costs. Colour lasers producing large-format output are also available for photo-realistic or artistic hardcopy. Until recently, the high cost ruled them out for the small design office, but increased production has brought price levels down very quickly. All lasers are simple to use and maintain, produce high quality output very fast and with great versatility and are ever-more likely to be used as an all-purpose output device in the design office. They are also clean in operation and need less maintenance than other types of device.

Pen plotters

These were invented for use with CAD drawings and mimic the movement of hand-held pens across drawing paper. This means that each individual line is drawn singly on paper or film and although the pens can accelerate and change direction quickly pen plotters are slower than some printer types which produce all he parts of each lines at every pass, such as inkjet devices.

The graphics processing of files is slowed down during the pen plotter method although the resolution of this form of output is very high, and accurate fine lines are easily produced. A wide range of media and pens can also be used to produce very varied presentation for different purposes.

Also, the pen plotter is not suitable for large areas of shading and rendering.

8.2 Pen plotter media

The pen plotter is an amazingly versatile device which can deal with many types of media using more types of applicator than any other device. In its form as a vinyl sign cutter it will continue to be used for cutting sheet materials to complex patterns long after the paper pen plotter is obsolete.

The description below is intended to illustrate the degree of versatility that can be achieved by the careful selection of media and pen types.

Cut sheet sizes A0–A5 can be used, as can continuous rolls of widths 24", 36" and 44" (the standard widths).

Hardcopy material

Translucent paper   For dependable check plots and quality originals for short-run dyelining at a competitive cost.

Bond or cartridge   Suitable for finished line drawings. When you need a quality appearance at an economical price, take a close look at bond and opaque paper. The whiteness and opacity of this media makes it perfect for copier use and photo-reproduction. The exceptionally smooth surface allows inking speeds without pen sticking, fibre build-up or feathering.

Polyester film   For long term master plots and archives. When you require the ultimate in accuracy – for example when you are plotting master archival copies or multi-layer schematics – then you need Polyester film media. These have extreme dimensional and chemical stability, even after long storage or in adverse environmental conditions.

How to choose the best media

Stability   Polyester film is your best bet.

Folding capability   100% rag, bond and translucent paper are all good choices.

Inking and line quality   Polyester film is the top recommendation, followed by vellum, 100% rag and natural tracing paper.

Contrast   100% rag and bond paper give excellent contrast.

Translucency   Polyester film and vellum are actually superior to the obvious choice of translucent.

Transparency   Polyester film is the best choice.

Economy   For low cost and high quality, translucent and bond paper provide the best economy.

Dyeline reproduction   Polyester film and vellum are the best choices.

Copier and photo reproduction   100% rag and bond are both excellent choices for this type of reproduction.

Archival quality   Polyester film is best, followed by 100% rag paper, for archiving purposes.

Speciality sizes   Translucent, natural tracing and bond paper have the widest range of available sizes.

Pens and pen types

Liquid ball   In various colours and line widths for accuracy and reliability.

Fibre tip   For bold thick lines (presentation work).

Plastic tip   For long life and wider lines.

Plotting media and pens combinations

Once you have selected the proper media for your application, in the correct format for your plotter the final step is to match up your media, pen and ink for the best possible performance from all the elements of your plotting system. Here is a small range of the many choices available:

Long-term archival storage   Polyester film and re-usable liquid ink tungsten tip pens with general purpose black ink.

Copier use and photo-reproduction   100% rag and liquid ink pens.

Dyelines and diazos   Vellum or natural tracing and liquid ink pens.

Contrast   Bond and plastic tip pens.

Check plotting   Translucent and liquid ball pens.

Final drawing   Vellum or natural tracing and liquid ink pens.

Overheads   Polyester film and plastic tip pens.

High speed   Translucent and liquid ball or pressurised ball pens.

Figure 8.1 indicates which papers and pens offer the best perform-ance combinations. Note that many other special types available for specialist uses.

Paper	Pen			
	Liquid Ball		Liquid Tip	
	Plastic	Fibre	Steel	Tungsten
Bond/cartridge	✓	✓	✓	✓
Tracing/vellum			✓	✓
Polyester film				✓
Translucent	✓	✓	✓	✓
Rag	✓		✓	✓

Fig. 8.1

8.3 Speciality plotters

Colour electrostatic plotters

These are used for producing large volumes of presentation drawings, artwork, maps and other top-quality work requiring a larger sheet size than thermal transfer plotters can accommodate. They work by electrostatic discharge which attracts liquid mate-rial to the media.

These very expensive machines provide high-quality, wide-format, dry-colour plots with superb output quality and a very high capacity; 400 dpi resolution is available for high precision copy. Users can define up to 1024 colours and 1024 lines per plot.

Some models accept information on of up to 256 layers, enabling the user to merge, overlay, sequentially overlay and erase data. This makes them very suitable for architectural and ground modelling systems which usually place entities on many layers.

The consumables for these machines require special handling as the chemicals involved can represent a health hazard.

Thermal transfer plotters

These are used for presentation drawings and artwork on smaller sheet sizes (A3 and A4). They uses rolls of different colour film which are transferred to the finished paper by heat action.

Thermal transfer plotters combined with colour video controllers can produce vibrant colour saturated A3- or A4-size drawings. Almost any workstation with RGB output capability can be connected to these plotters.

Video memories of 2–4 Mb are needed for high-resolution graphics cards used with monitors capable of 1280×1024 pixel resolution with rapid screen-capture times of one second or less. The printing process combines a thermal imaging head with thermally sensitive ink ribbon to transfer dots of yellow, magenta or cyan on to both paper and transparency film as required. Various resolutions are available.

The three primary colours produce red, green and blue, these plus yellow, magenta, cyan and black give seven solid colours available for any single dot. Through sophisticated dithering techniques thermal transfer printers can output up to 4096 colours. Some plotters offer three or more levels of colour intensity. Several workstations can usually be connected to one plotter.

Other features include:

• black and white reversal
• fractional enlargement
• multiple copy generation

These plotters offer easy set up and operation and smaller cheaper machines are not available but as they have limited memory speed of plotting is not startling.

8.4 Printing and plotting routines

The Plot command selected from the File pull down menu or from the tablet [W24] is used to access a large multi-function dialogue box that controls all facets of hardcopy output.

Each project described in this book, just as in real-life project work, calls for a different type of hardcopy; the routines listed

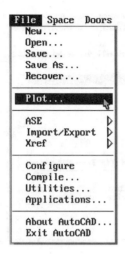

Fig. 8.2

below are those that could be used for a typical print or plotting routine.

Practical

During this session you will practise the following:

- Selecting an output device for the hardcopy.
- Choosing a paper size and orientation.
- Previewing the proposed plot.
- Plotting the drawing file.
- Examining alternative output methods.
- The configuration of AutoCAD plotters and other devices.

Device selection

First select the device to be used for plotting:

1   Pick Plot from the File pull-down menu or [W24] from the tablet, as shown in Fig. 8.2.
2   Click on Device and default selection ... in the Plot configuration dialogue box, as shown in Fig. 8.3.
4   Click on one of the devices available in the Device and default selection dialogue box, as shown in Fig. 8.4.
    Comment   The hardcopy devices available will be limited by the number of ports connected to your computer but will usually include two parallel ports (designated 'LPT') and two serial ports ('COM').
4   Click on OK to confirm selection and return to the previous plot configuration dialogue box.

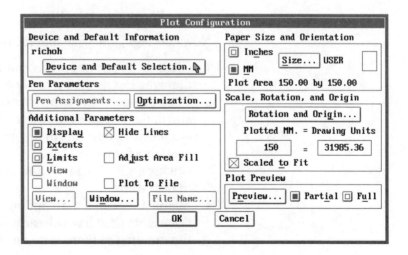

Fig. 8.3

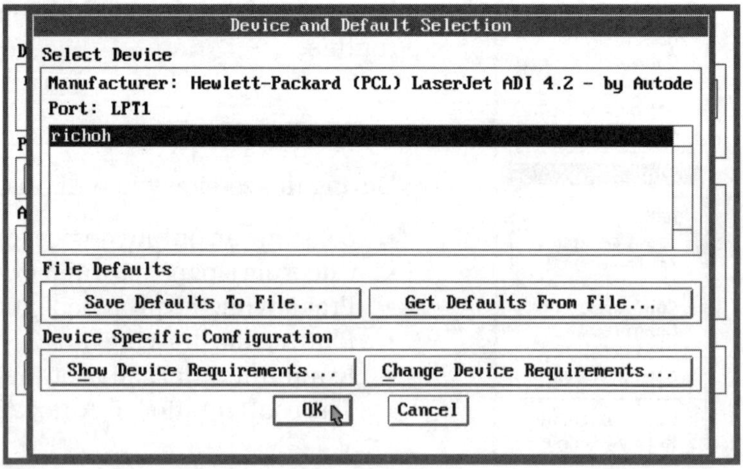

Fig. 8.4

Outputting as an EPS file

Instead of directly printing a drawing, an EPS 'printer file' can be selected as the output device; this method creates a Postscript printing file which can be then output later on any Postcript device.

4   Select Import/export on the File menu or [W25] on the tablet.
5   Select the Plot to file option and pick File name ... in the Plot configuration dialogue box.
6   Type in a suitable file name in the Create Postscript file dialogue box.

Once the file has been created, you can print the file on any Postscript device, now or later.

Paper size selection

Now choose a paper size that the device can handle.

1   Click on Paper size and orientation in the plot configuration dialogue box as shown in Fig. 8.3.
2   Click on one of the paper size options that appears in the Paper size dialogue box as shown in Fig. 8.5.
    Comment   Note the dual imperial and metric units facility available for paper sizes.
3   Click on OK to select and return to the Plot Configuration dialogue box.
4   Enable the Scaled to fit option.
    Comment   You have already set up the scale and sheet size for the drawing so there is no need to do it again here. If you wish

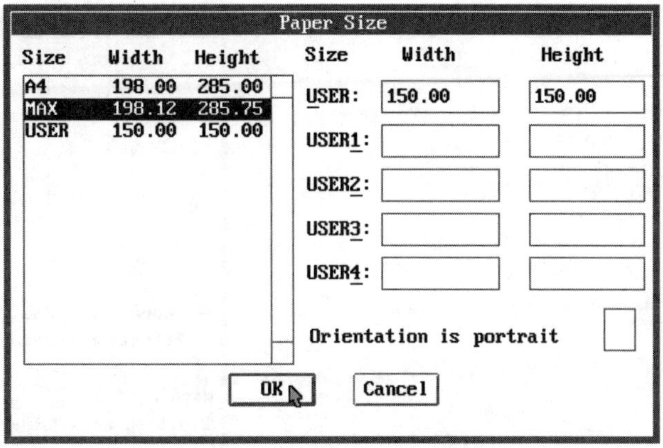

Fig. 8.5

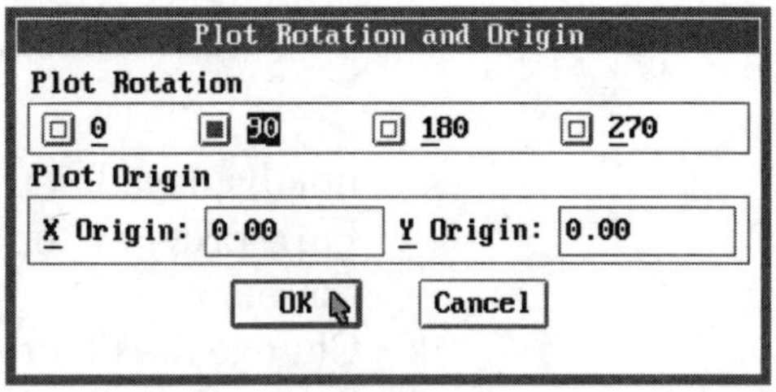

Fig. 8.6

to change the plot origin or rotation angle (for insertion in a book or other document) click on the Plot rotation and origin box as shown in Fig. 8.6 and select 90 degrees.

5   Enable Display in the Additional parameters section of the Plot configuration dialogue box to plot the display on screen, as shown in Fig. 8.3.

6   The other options determining what is plotted are:

Display = what is shown on the monitor
Extents = the full extent of the drawing
Limits = the coordinates input determines the amount of the drawing plotted
View = any previously saved view with a file name
Window = area windowed on the screen

Plot preview

Now preview the plot to make sure that it will fit the paper selected using either the partial or full preview options.

1   Enable Partial plot preview and click on Preview ... in the Plot configuration dialogue box.
Comment   As you can see in Fig. 8.7 a warning about the plotting area being truncated has appeared. The full drawing is

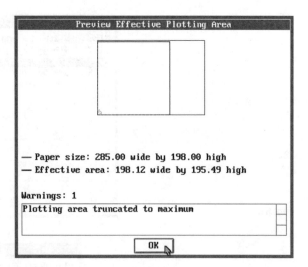

Fig. 8.7

not displayed in this view to save time and you must enable the Full preview option to regenerate these drawing files on screen.

2   Change the Plot rotation back to 0 degrees.

3   Select Full, Preview ... from the Plot configuration dialogue box.

   Comment   The full drawing will appear on screen as shown in Fig. 8.8.

4   Select Pan and zoom to reposition the drawing on screen or, if

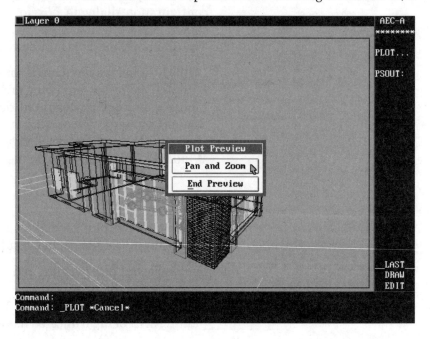

Fig. 8.8

happy with the display, select End preview prior to plotting as shown in Fig. 8.8.

The final step depends on your hardcopy device setup but it is normally only necessary to hit *enter* to print or plot.

Alternative hardcopy production

To explore the alternative methods of producing hardcopy output you must unfortunately change to the AutoCAD menu and then use the rendering facilities found there.

1 Pick the Acad.mnu option from the main menu on the right hand side of the screen.
2 Pick Save image … from the Render pull-down menu as shown in Fig. 8.9.
3 Select TGA as the file format from the dialogue box, as shown in Fig. 8.10.
   Comment   In this case you are saving in TGA format for export to a graphics enhancement programme.
4 Type in a file name of your choice in the Image name box, choosing a directory within which to save the file.
5 Choose Render from the Render pull down menu. The rendered image will appear on screen as shown in Fig. 8.11.
   Comment   This is a default render, without setting up any lights or scenes, just to illustrate another way of producing alternative file or hardcopy output. The reader should consult the AutoCAD Release 12 reference manual for for full details of

Fig. 8.9

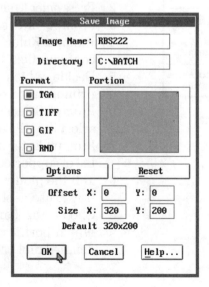

Fig. 8.10

Fig. 8.11

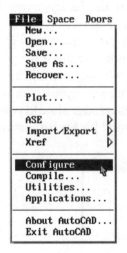

Fig. 8.12

how to set up a 3D model for more effective rendered output.

If you need to set up an alternative device for printing you must tell AEC that you have connected it by reconfiguring AutoCAD.

Reconfiguring AutoCAD

1   Pick Configure from the File pull-down menu as shown in Fig. 8.12.
    Comment   The current AutoCAD configuration will be displayed as shown in Fig. 8.13.
2   Press *enter* to show the configuration menu on screen.
3   Prompt: Enter selection <0>: **5** *enter*, as shown in Fig. 8.14.
4   Prompt: Enter selection 0 to 4 <0>: **1** *enter* to add a plotter configuration.
5   Decide which device you want to add from the available printers listed or *enter* to see more as shown in Fig. 8.15.
6   Prompt: Select device number or ? to repeat list <1>: **12** *enter* to select the HP Laserjet as shown in Fig. 8.16.
7   Prompt: Enter selection: decide which model to pick from the list of supported models.
8   Tell the plotter which port it is connected to and which paper tray to use, as shown in Fig. 8.16.
9   Choose the hardcopy resolution, number of hardcopy plots, which port to connect and if you want to change any of the detailed plot parameters, all as shown in Fig. 8.17.

Fig. 8.13

```
Current AutoCAD configuration


  Video display:      IBM Video Graphics Array ADI 4.2 - by Autodesk

IBM VGA v1.8d (21jun92).  Universal Super VGA ADI (Display/Render)
DOS Protected Mode ADI 4.2 Driver for AutoCAD.
Config file is C:\ACAD12\DRV\SVADI.CFG
Configured for: IBM Video Graphics Array.  Text font: 8x16.
Display - 640x480 in 16 colors on Light background.
Rendering - 320x200 in 256 colors.

    Version: A.1.18

  Digitizer:          Microsoft Mouse Driver ADI 4.2 - by Autodesk
                      Microsoft Mouse
    Version: A.1.18

  Plotter:            Hewlett-Packard (PCL) LaserJet ADI 4.2 - by Autodesk
                      HP LaserJet III
    Port: Parallel Printer LPT1 at address 378 (hex)
    Version: A.1.18

-- Press RETURN for more --
```

Fig. 8.14

```
Configuration menu

  0.  Exit to drawing editor
  1.  Show current configuration
  2.  Allow detailed configuration

  3.  Configure video display
  4.  Configure digitizer
  5.  Configure plotter
  6.  Configure system console
  7.  Configure operating parameters

Enter selection <0>: 5
Your current plotter is: Hewlett-Packard (PCL) LaserJet ADI 4.2 - by Autodesk
Description: richoh

Plotter Configuration Menu

  0.  Exit to configuration menu
  1.  Add a plotter configuration
  2.  Delete a plotter configuration
  3.  Change a plotter configuration
  4.  Rename a plotter configuration

Enter selection, 0 to 4 <0>:
```

Fig. 8.15

```
Obtained from:  AUTODESK.LTD - - - - - -

Available plotters:

  1.  None
  2.  ADI plotter or printer (installed - pre v4.1) - by Autodesk
  3.  AutoCAD file output formats (pre 4.1) - by Autodesk
  4.  CalComp ColorMaster Plotters ADI 4.2 - by Autodesk
  5.  CalComp DrawingMaster Plotters ADI 4.2 - by Autodesk
  6.  CalComp Electrostatic Plotters ADI 4.2 - by Autodesk
  7.  CalComp Pen Plotters ADI 4.2 - by Autodesk
  8.  Canon Laser Printer ADI 4.2 - by Autodesk
  9.  Epson printers ADI 4.2 - by Autodesk
  10.  Hewlett-Packard (HP-GL) ADI 4.2 - by Autodesk
  11.  Hewlett-Packard (HP-GL/2) ADI 4.2 - by Autodesk
  12.  Hewlett-Packard (PCL) LaserJet ADI 4.2 - by Autodesk
  13.  Hewlett-Packard (PCL) PaintJet XL ADI 4.2 - by Autodesk
  14.  Houston Instrument ADI 4.2 - by Autodesk
  15.  IBM 7300 Series ADI 4.2 - by Autodesk
  16.  IBM Graphics Printer <obsolete> ADI 4.2 - by Autodesk
  17.  IBM Proprinter ADI 4.2 - by Autodesk
  18.  JDL 750 & 750E <obsolete> ADI 4.2 - by Autodesk
  19.  NEC Pinwriter P5/P5XL/P9XL <obsolete> ADI 4.2 - by Autodesk
  20.  PostScript device ADI 4.2 - by Autodesk
-- Press RETURN for more --
```

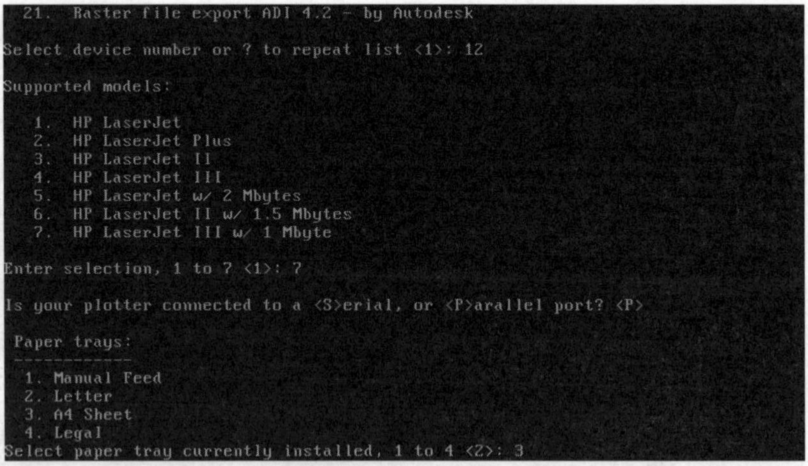

Fig. 8.16

```
  21.   Raster file export ADI 4.2 - by Autodesk

Select device number or ? to repeat list <1>: 12

Supported models:

    1.   HP LaserJet
    2.   HP LaserJet Plus
    3.   HP LaserJet II
    4.   HP LaserJet III
    5.   HP LaserJet w/ 2 Mbytes
    6.   HP LaserJet II w/ 1.5 Mbytes
    7.   HP LaserJet III w/ 1 Mbyte

Enter selection, 1 to 7 <1>: 7

Is your plotter connected to a <S>erial, or <P>arallel port? <P>

Paper trays:
------------

    1.   Manual Feed
    2.   Letter
    3.   A4 Sheet
    4.   Legal
Select paper tray currently installed, 1 to 4 <2>: 3
```

Fig. 8.17

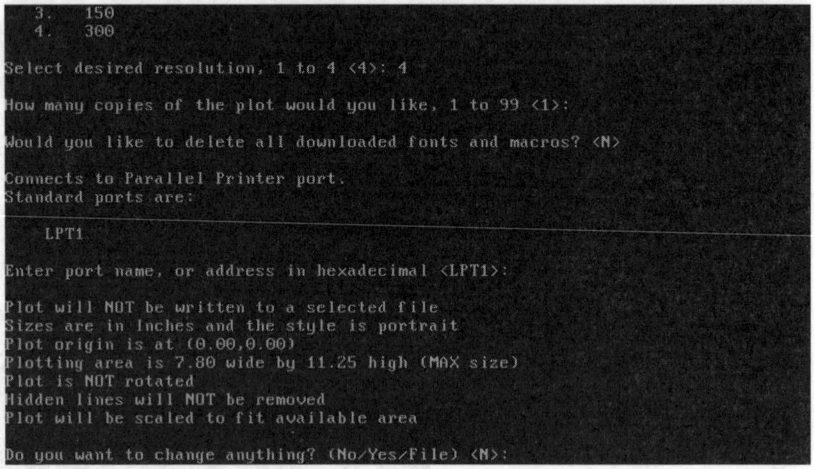

```
    3.    150
    4.    300

Select desired resolution, 1 to 4 <4>: 4

How many copies of the plot would you like, 1 to 99 <1>:

Would you like to delete all downloaded fonts and macros? <N>

Connects to Parallel Printer port.
Standard ports are:

    LPT1

Enter port name, or address in hexadecimal <LPT1>:

Plot will NOT be written to a selected file
Sizes are in Inches and the style is portrait
Plot origin is at (0.00,0.00)
Plotting area is 7.80 wide by 11.25 high (MAX size)
Plot is NOT rotated
Hidden lines will NOT be removed
Plot will be scaled to fit available area

Do you want to change anything? (No/Yes/File) <N>:
```

Raster plot files

As an alternative to direct plotter or printer output the drawing can be written to a plot file for importation into another software package or for plotting later in a different file format. This is a particularly useful feature when the drawing forms only a part of a compound graphics document which may involve using a photographic background. The production of photo-realistic presentation artwork that involves the merging of CAD and photo-realistic images can now be achieved on a 486-class PC with 16–32 Mb of RAM. The first stage in this process is to write a plot file in a raster format that the artwork package can understand. Auto-CAD can produce raster files in all the most commonly used formats and a typical procedure for creating a TGA plot file is described below.

```
 1.   None
 2.   ADI plotter or printer (installed - pre v4.1) - by Autodesk
 3.   AutoCAD file output formats (pre 4.1) - by Autodesk
 4.   CalComp ColorMaster Plotters ADI 4.2 - by Autodesk
 5.   CalComp DrawingMaster Plotters ADI 4.2 - by Autodesk
 6.   CalComp Electrostatic Plotters ADI 4.2 - by Autodesk
 7.   CalComp Pen Plotters ADI 4.2 - by Autodesk
 8.   Canon Laser Printer ADI 4.2 - by Autodesk
 9.   Epson printers ADI 4.2 - by Autodesk
10.   Hewlett-Packard (HP-GL) ADI 4.2 - by Autodesk
11.   Hewlett-Packard (HP-GL/2) ADI 4.2 - by Autodesk
12.   Hewlett-Packard (PCL) LaserJet ADI 4.2 - by Autodesk
13.   Hewlett-Packard (PCL) PaintJet XL ADI 4.2 - by Autodesk
14.   Houston Instrument ADI 4.2 - by Autodesk
15.   IBM 7300 Series ADI 4.2 - by Autodesk
16.   IBM Graphics Printer <obsolete> ADI 4.2 - by Autodesk
17.   IBM Proprinter ADI 4.2 - by Autodesk
18.   JDL 750 & 750E <obsolete> ADI 4.2 - by Autodesk
19.   NEC Pinwriter P5/P5XL/P9XL <obsolete> ADI 4.2 - by Autodesk
20.   PostScript device ADI 4.2 - by Autodesk
-- Press RETURN for more --
21.   Raster file export ADI 4.2 - by Autodesk

Select device number or ? to repeat list <1>: 21
```

Fig. 8.18

```
 7.   800 x 600
 8.   1024 x 768
 9.   1152 x 900   (Sun standard)
10.   1600 x 1280  (Sun hi-res)
11.   User-defined

Enter selection, 1 to 11 <1>: 8

You can export the drawing in any of the following raster file
formats.  Please select the format you prefer.

 1.   GIF (CompuServe Graphics Interchange Format)
 2.   X Window dump (xwd compatible)
 3.   Jef Poskanzer's Portable Bitmap Toolkit Formats
 4.   Microsoft Windows Device-independent Bitmap (.BMP)
 5.   TrueVision TGA Format
 6.   Z-Soft PCX Format
 7.   Sun Rasterfile
 8.   Flexible Image Transfer System (FITS)
 9.   PostScript image
10.   TIFF (Tag Image File Format)
11.   FAX Image (Group 3 Encoding)
12.   Amiga IFF / ILBM Format

In which format would you like to export the file, 1 to 12 <1>: 5
```

Fig. 8.19

If you need to set up an alternative device for writing a plot file you must tell AEC that you have added that option by reconfiguring AutoCAD.

1 Pick Configure from the File pull-down menu.
   Comment   The current AutoCAD configuration will be displayed.
2 Press *enter* to show the configuration menu on screen.
3 Prompt: Enter selection <0>: **5** *enter*.
4 Prompt: Enter selection 0 to 4 <0>: **1** *enter* to add a plotter configuration.
5 Prompt: Select device number or ? to repeat list <1>: **21** *enter* as shown in Fig. 8.18.
6 Select a screen resolution, from the options listed in Fig. 8.19, to

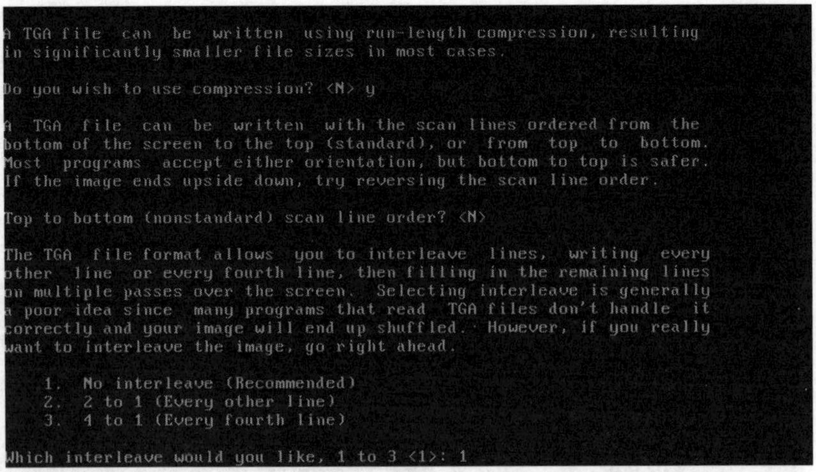

A TGA file can be written using run-length compression, resulting
in significantly smaller file sizes in most cases.

Do you wish to use compression? <N> y

A TGA file can be written with the scan lines ordered from the
bottom of the screen to the top (standard), or from top to bottom.
Most programs accept either orientation, but bottom to top is safer.
If the image ends upside down, try reversing the scan line order.

Top to bottom (nonstandard) scan line order? <N>

The TGA file format allows you to interleave lines, writing every
other line or every fourth line, then filling in the remaining lines
on multiple passes over the screen. Selecting interleave is generally
a poor idea since many programs that read TGA files don't handle it
correctly and your image will end up shuffled. However, if you really
want to interleave the image, go right ahead.

    1. No interleave (Recommended)
    2. 2 to 1 (Every other line)
    3. 4 to 1 (Every fourth line)

Which interleave would you like, 1 to 3 <1>: 1

Fig. 8.20

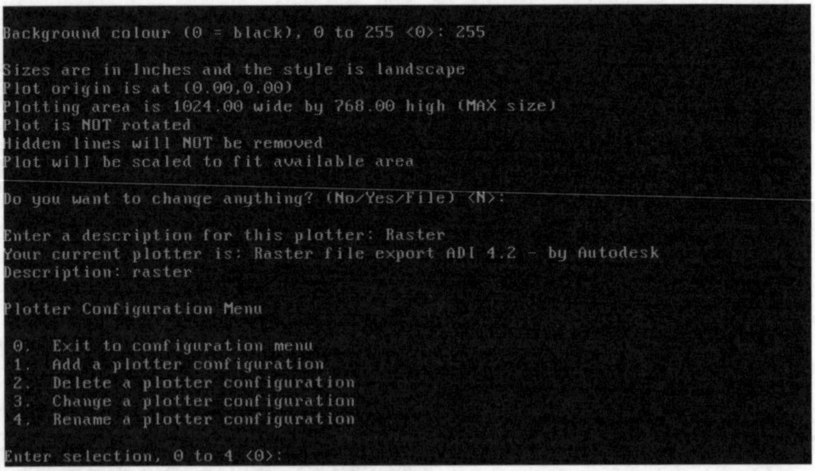

Background colour (0 = black), 0 to 255 <0>: 255

Sizes are in Inches and the style is landscape
Plot origin is at (0.00,0.00)
Plotting area is 1024.00 wide by 768.00 high (MAX size)
Plot is NOT rotated
Hidden lines will NOT be removed
Plot will be scaled to fit available area

Do you want to change anything? (No/Yes/File) <N>:

Enter a description for this plotter: Raster
Your current plotter is: Raster file export ADI 4.2 - by Autodesk
Description: raster

Plotter Configuration Menu

0. Exit to configuration menu
1. Add a plotter configuration
2. Delete a plotter configuration
3. Change a plotter configuration
4. Rename a plotter configuration

Enter selection, 0 to 4 <0>:

Fig. 8.21

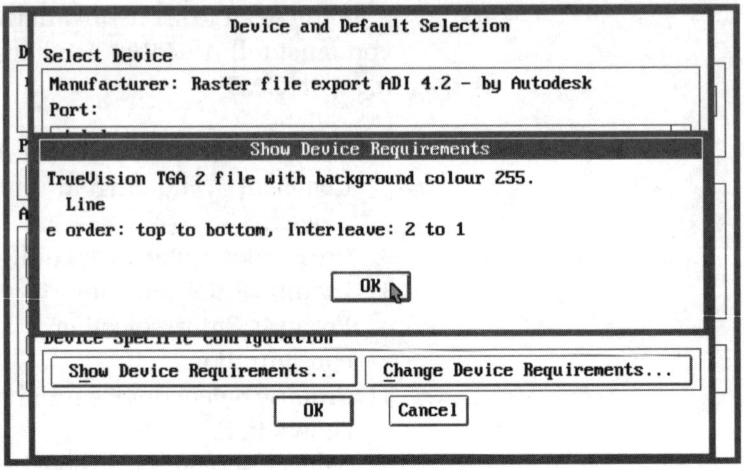

Device and Default Selection

Select Device

Manufacturer: Raster file export ADI 4.2 - by Autodesk
Port:

Show Device Requirements

TrueVision TGA 2 file with background colour 255.
    Line
e order: top to bottom, Interleave: 2 to 1

OK

Device Specific Configuration

Show Device Requirements...   Change Device Requirements...

OK   Cancel

Fig. 8.22

suit your graphics card and monitor; here the correct option is 8: 1024×768 dpi.

7 Decide which type of file format you want to create; here the required option is the TGA format option (number 5), as shown in Fig. 8.19.
8 Choose the other options as shown in Figs 8.20 and 8.21 to complete the process of adding a plotter device and be sure to save the configuration.

Note File compression reduces the graphics file size. The Auto-CAD text gives advice about possible optional selections.

Next use the added plotting device to create a plot file in TGA format.

8 Pick Plot from the File menu.
9 Pick Device and default selection from the Plot configuration dialogue box and select the Raster file export ADI 4.2 option.
10 Click on Show device requirements ... to confirm that a TrueVision TGA 2 file with a white background will be written as shown in Fig. 8.22.

Note Note that the programme has reset the interleave value.

11 Click on OK and then on File name in the Plot configuration menu.
12 Enter a file name for the plot file in the dialogue box and click on OK.
13 Check that the Plot to file option is enabled.

The plot file will now be written to the hard disk.

Index